CHEMISTRY RESEARCH AND APPLICATIONS

DIPICOLINIC ACID, ITS ANALOGUES, AND DERIVATIVES: ASPECTS OF THEIR COORDINATION CHEMISTRY

CHEMISTRY RESEARCH AND APPLICATIONS

Additional books in this series can be found on Nova's website under the Series tab.

CHEMISTRY RESEARCH AND APPLICATIONS

DIPICOLINIC ACID, ITS ANALOGUES, AND DERIVATIVES: ASPECTS OF THEIR COORDINATION CHEMISTRY

ALVIN A. HOLDER
LESLEY C. LEWIS-ALLEYNE
DON VANDERVEER
AND
MARVADEEN SINGH-WILMOT

Nova Science Publishers, Inc.
New York

Copyright ©2011 by Nova Science Publishers, Inc.

All rights reserved. No part of this book may be reproduced, stored in a retrieval system or transmitted in any form or by any means: electronic, electrostatic, magnetic, tape, mechanical photocopying, recording or otherwise without the written permission of the Publisher.

For permission to use material from this book please contact us:
Telephone 631-231-7269; Fax 631-231-8175
Web Site: http://www.novapublishers.com

NOTICE TO THE READER

The Publisher has taken reasonable care in the preparation of this book, but makes no expressed or implied warranty of any kind and assumes no responsibility for any errors or omissions. No liability is assumed for incidental or consequential damages in connection with or arising out of information contained in this book. The Publisher shall not be liable for any special, consequential, or exemplary damages resulting, in whole or in part, from the readers' use of, or reliance upon, this material. Any parts of this book based on government reports are so indicated and copyright is claimed for those parts to the extent applicable to compilations of such works.

Independent verification should be sought for any data, advice or recommendations contained in this book. In addition, no responsibility is assumed by the publisher for any injury and/or damage to persons or property arising from any methods, products, instructions, ideas or otherwise contained in this publication.

This publication is designed to provide accurate and authoritative information with regard to the subject matter covered herein. It is sold with the clear understanding that the Publisher is not engaged in rendering legal or any other professional services. If legal or any other expert assistance is required, the services of a competent person should be sought. FROM A DECLARATION OF PARTICIPANTS JOINTLY ADOPTED BY A COMMITTEE OF THE AMERICAN BAR ASSOCIATION AND A COMMITTEE OF PUBLISHERS.

Additional color graphics may be available in the e-book version of this book.

LIBRARY OF CONGRESS CATALOGING-IN-PUBLICATION DATA

Dipicolinic acid, its analogues and derivatives : aspects of their coordination chemistry / authors, Alvin A. Holder ... [et al.].
    p. cm.
  Includes bibliographical references and index.
  ISBN 978-1-61209-770-1 (softcover)
  1. Carboxylic acids. 2. Pyridine--Derivatives. I. Holder, Alvin A.
  QD305.A2D64 2011
  547'.593--dc23
                       2011015367

Published by Nova Science Publishers, Inc. † New York

CONTENTS

Preface		vii
Chapter 1	Introduction	1
Chapter 2	Coordination Chemistry of Dipicolinic Acid and Its Analogues	5
Chapter 3	Conclusions	65
References		67
Index		77

PREFACE

2,6-Pyridinedicarboxylic acid (dipicolinic acid) is a widely used building block in coordination and supramolecular chemistry. The crystal structure of dipicolinic acid was first solved in 1973, which confirmed its molecular formula of $C_7H_5NO_4$, a molar mass of 167.119 g mol^{-1}, and the resulting composition of its constituent atoms (C, 50.31%; H, 3.02%; N, 8.38%; and O, 38.29%). Dipicolinic acid and its analogues are known to form many intriguing complexes with main group and other metal ions from as far back as 1877. The corresponding bis-acid (DPA), bis-ester (DPE), and bis-amide (DPAM) derivatives behave as tridentate ligands, which efficiently coordinate to various metal ions. This book will discuss the coordination chemistry of several metal complexes with dipicolinic acid, its analogues, and derivatives as ligands.

Chapter 1

INTRODUCTION

2,6-Pyridinedicarboxylic acid (dipicolinic acid), **I**, is a widely used building block in coordination and supramolecular chemistry.[1-5] It is a versatile, strong, nitrogen-oxygen, multi-modal donor ligand, which forms stable complexes with diverse metal ions, sometimes in unusual oxidation states, for example, its corresponding bis-acid (DPA), bis-ester (DPE), and bis-amide (DPAM) derivatives behave as tridentate ligands, which efficiently coordinate to various metal ions.[6-13]

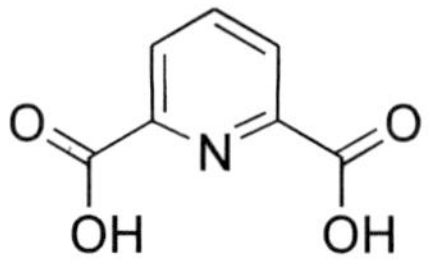

I

The crystal structure of dipicolinic acid was first solved by Takusagawa *et al.*[14] in 1973. Takusagawa *et al.*[14] confirmed its molecular formula of $C_7H_5NO_4$, a molar mass of 167.119 g mol^{-1}, and the resulting composition of its constituent atoms (C, 50.31%; H, 3.02%; N, 8.38%; and O, 38.29%).

More recently, (creatH)$^+$(Hdipic)$^-$.H$_2$O was synthesized by the reaction between dipicolinic acid and creatinine (creat) (Scheme 1).[15] Its structure consists of (creatH)$^+$ and (Hdipic)$^-$ ions and a disordered water molecule (Figure 1), all lying on a crystallographic mirror plane.[15] It was reported that the intermolecular interactions among these three fragments consist of ion-pairing, hydrogen bonding and π-π stacking. A single proton transfer occurs from one of the two carboxylic acid functional groups to the endocyclic imine N atom of creatinine. This results in the localization of the exocyclic C8-N4 double bond [1.300 (2) Å] and the adjacent single bond C8-N3 [1.369 (2) Å]. These values can be compared with the intermediate, delocalized values in the parent neutral creatinine molecule [1.320 (3) and 1.349 (3) Å, respectively].[16] The two carboxylic groups of the (Hdipic)⁻ anion adopt slightly different conformations, both being essentially coplanar with the pyridine ring. It was reported that all of the N and O heteroatoms participate in extensive strong or weak hydrogen-bonding interactions, particularly the strong O3•••O2^i interaction.

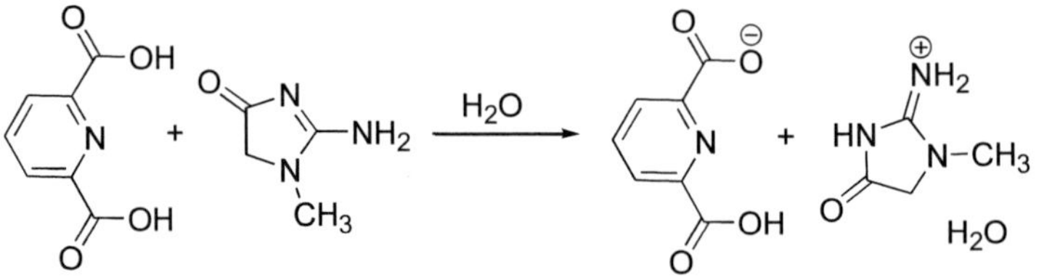

Scheme 1.

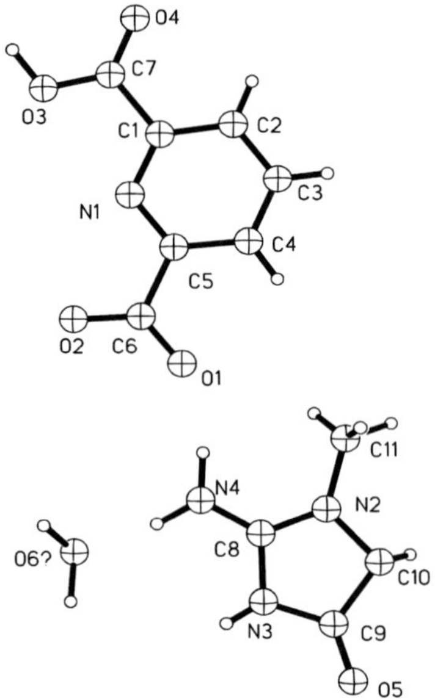

Figure 1. A diagram of (creatH)$^+$(Hdipic)⁻·H$_2$O. (Reproduced by permission from reference 15).

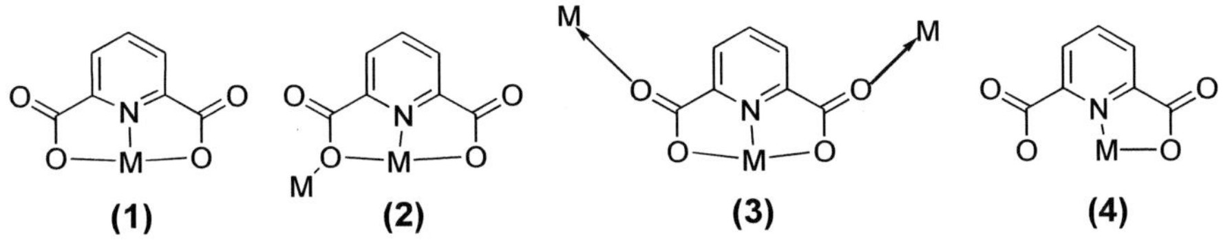

Figure 2. Different coordination modes of the dipicolinate anion.

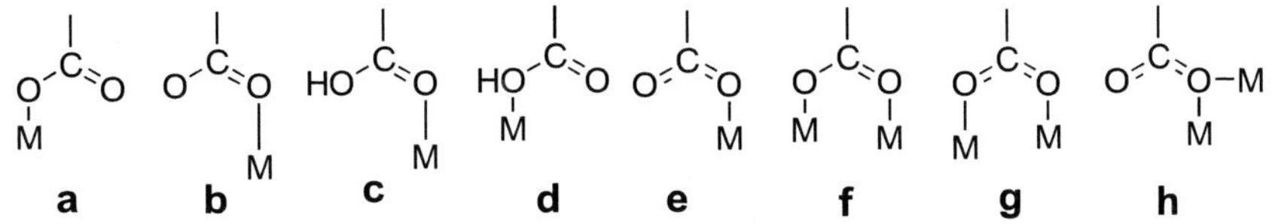

Figure 3. Coordination modes observed in the solid state of metal-carboxylate complexes.

Dipicolinic acid and its analogues are known to form many intriguing complexes with main group and transition metal ions from as far back as 1877.[17] Since the discovery of the dipic^{2-} ligand in a biological system,[18] its coordination chemistry has been extensively investigated. Several modes of coordination are known: O-unidentate[19] H$_2$dipic; O,N-bidentate[20] Hdipic$^-$; O,N,O-tridentate[19, 21-26] H$_2$dipicH, Hdipic$^-$, and dipic^{2-}; (O,N)O-bidentate bridging[27, 28] dipic^{2-}; (O,N,O)O and O,N,i-O-tridentate bridging[23, 25, 29-33] dipic^{2-}. There has been renewed interest in the complexes of this ligand from several standpoints including unconventional physical properties such as liquid crystal behavior and nonlinear optics,[26, 28] DNA cleavage,[34] electron transfer,[35-39] activation of dioxygen,[40-48] and novel coordination modes.[28]

Different coordination modes have been reported in transition metal-dipicolinate complexes.[28] These are shown in Figure 2:

The coordination modes observed in the solid state of metal-carboxylate complexes are shown in Figure 3. The coordination modes previously reported are labeled a,[23, 25, 49-54] c, [22, 23, 25, 49-55] e,[22, 52, 54] f,[23] g,[51] and h.[25] Coordination modes b and d have been reported in the literature.[56]

The most common coordination mode is **a** in which the metal is coordinated to the long C-O bond.[56] This coordination mode is found in [Ni(Hdipic)$_2$].3H$_2$O,[49, 50, 53] [Zn(Hdipic)$_2$].3H$_2$O,[51, 52] [Zn$_2$(dipic)$_2$

].7H$_2$O,[51] [Fe(Hdipic)$_2$(OH$_2$)],[25] [Fe$_2$(dipic)$_2$(OH$_2$)$_6$]. 2H$_2$dipic, [25] [Fe$_3$(dipic)$_2$ (Hdipic)$_2$ (H$_2$O)$_4$].2H$_2$O,[23] [Fe$_2$ (dipic)$_2$ (H$_2$O)$_5$] .2.25H$_2$O ,[23] [Fe$_3$(dipic)$_4$ (H$_2$O)$_6$ (NH$_4$)$_2$].4 H$_2$O. 2H$_2$dipic,[23] [Fe$_{13}$ (Hdipic) $_6$(dipic)$_{10}$ (H$_2$O)$_{24}$]. 13H$_2$O ,[23] [Cu(dipic) (H$_2$O)$_2$],[54] and [Cu(H$_2$d ipic)(d ipic)].H$_2$O.[52] Another common coordination mode found in complexes with monoprotonated Hdipic⁻ is represented by **c** and found in [Ni (Hdipic)$_2$].3H$_2$O,[49, 50, 53] [Zn(Hdipic)$_2$]. 3H$_2$O,[51, 52] [Fe (Hdipic)$_2$ (OH$_2$)], 20 [Fe$_3$ (dipic)$_2$ (Hdipic)$_2$(H$_2$O)$_4$].2H$_2$O,[23] [Fe$_{13}$(Hdipic)$_6$(dipic)$_{10}$(H$_2$O)$_{24}$].13H$_2$O, [23] and [Cu(H$_2$dipic)(dipic)].H$_2$O.[52, 55] A coordination mode in which the C-O bonds are of equal length is **e**, which has been observed in [Ag(Hdipic)$_2$]. H$_2$O,[22] [Cu(H$_2$dipic)(dipic)].3H$_2$O,[52, 55] and [Zn(Hdipic)$_2$]. 3H$_2$O.[52] Less commonly observed coordination modes involve coordination of two metal ions to one carboxylate group and are represented by **f** (observed in [Fe$_3$(dipic)$_2$ (Hdipic)$_2$ (H$_2$O)$_4$] .2H$_2$O, [23] [Fe$_2$ (dipic)$_2$ (H$_2$O)$_5$].2.25 H$_2$O,[23] and [Fe$_{13}$(Hdipic)$_6$(dipic)$_{10}$ (H$_2$O)$_{24}$].13H$_2$O),[23] **g** (observed in [Zn$_2$ (dipic)$_2$]. 7H$_2$O),[51] and **h** (observed in [Fe$_2$ (dipic)$_2$ (OH$_2$)$_6$]. 2H$_2$dipic).[25] The key differences in the two new coordination modes are the coordination of the metal ion to the short C=O bond (in **b**) and to the oxygen atom carrying the proton (in **d**).

More recently, there was a report on the use of dipicolinic acid in the design of layered crystalline materials using coordination chemistry and hydrogen bonds. MacDonald *et al.*[57] reported the synthesis and characterization of several first-row transition metals with dipicolinic acid as a ligand. Five bis(imidazolium 2,6-pyridinedicarboxylate)M(II) trihydrate complexes (where M = Mn^{2+}, Co^{2+}, Ni^{2+}, Cu^{2+}, or Zn^{2+}), were synthesized from the reaction between dipicolinic acid and imidazole with Mn^{2+}, Co^{2+}, Ni^{2+}, Cu^{2+}, or Zn^{2+} salts.[57]

Chapter 2

COORDINATION CHEMISTRY OF DIPICOLINIC ACID AND ITS ANALOGUES

This chapter discusses the coordination chemistry of selected main group and transition metal complexes with dipicolinic acid, its analogues, and derivatives as ligands. Selected elements will be presented in terms of increasing atomic number. Out of all of the alkali metals, there has been a report of the crystal structure of sodium coordinated to dipicolinic acid.[58] Calcium, magnesium, and strontium, three alkaline earth metals, are popular metal centers, which have been reported in the literature to be coordinated to dipicolinic acid or its analogues.[32, 33, 59-62]

The structure of the deep red diaquoperoxotitanium(IV) dipicolinate complex, $[TiO_2(C_7H_3O_4N)(H_2O)_2]\cdot 2H_2O$ was reported.[63] The complex (see Figure 4) has a pentagonal bipyramidal seven-fold coordination with two carboxylate oxygens, one nitrogen and two oxygens of the peroxo group forming a distorted pentagon and two water oxygens at the apices. The peroxo group is attached laterally to the titanium(IV) metal center. It was reported that the pentagon is virtually planar, with the distances of Ti from the least square plane being less than standard deviation.[63]

It was reported that the O-O distance in the peroxide is 1.458 Å, and that this value agreed well with values of 1.464, 1.463, and 1.469 Å in the triclinic diaquo, difluoro, and nitrilotriacetic acid (NTA) complexes, respectively.[63] The Ti-O_{peroxo} distance (1.833 Å) was compared with the values 1.834 and 1.856 Å, 1.846 and 1.861 Å, and 1.889 and 1.892 Å, respectively, and the Ti-O_{water} apical distances of 2.018 Å with the Ti-O (F) values of 2.022 and 2.055 Å in the triclinic diaquo, 1.853 and 1.887 Å in the difluoro and 1.819 and 2.065 Å in the NTA complexes.[63] It was concluded that while the O-O bond distance of the peroxo group was practically the same in all the structures, there was a small but significant variation in the Ti-O$_{peroxo}$ and apical distances.[63] There was progressive increase in the Ti-O$_{peroxo}$ bond lengths down the respective series, and a corresponding decrease in the apical bond lengths.[63]

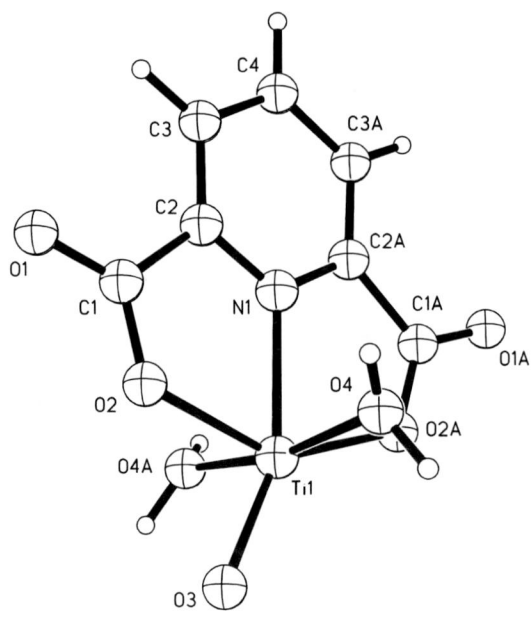

Figure 4. A diagram of [TiO$_2$(C$_7$H$_3$O$_4$N)(H$_2$O)$_2$].2H$_2$O. (Reproduced by permission from reference 63).

Reaction of (C$_5$H$_5$)$_2$Ti(CH$_3$)$_2$ or (CH$_3$)$_4$C$_2$(C$_5$H$_4$)$_2$Ti(CH$_3$)$_2$ with dipicolinic acid produced several titanocene dipicolinate derivatives.[64] Figure 5 shows the structure of one of those derivatives. As expected from structural studies on other transition metal dipicolinate complexes,[19, 21-26] the dipicolinate ligand is bound to the Ti(IV) metal centre by its pyridine N atom and two of the carboxylate O atoms, which occupy the central and the two lateral coordination sites of the titanocene fragment. The Ti-N and Ti-O distances of 216 and 211 ppm were reported to be significantly longer than Ti-N≡ bonds (196-202 pm)[65-67] and Ti-O bonds (186-190 pm)[68-70] in comparable, tetracoordinate titanocene complexes. The O-Ti-N angle of 71.1° was reported to be within the range of 65-73° found in other pentacoordinate, non-hydridic metallocene derivatives.[71, 72]

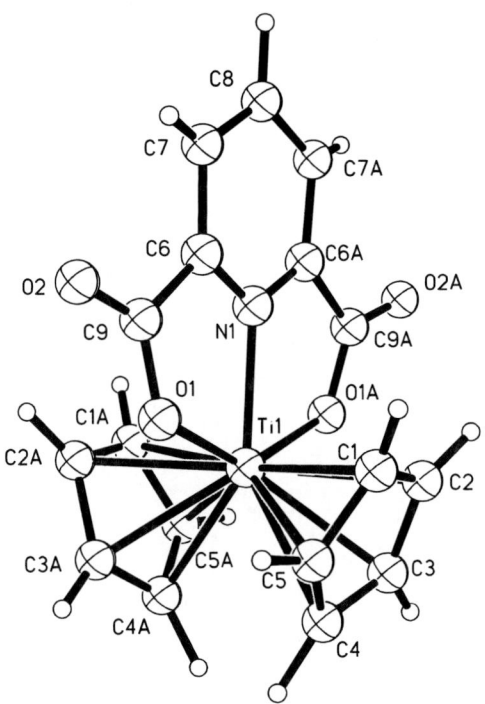

Figure 5. A diagram of [(C₅H₅)₂Ti(dipic)]. (Reproduced by permission from reference 64).

The titanium(IV) metal center and its N an O ligand atoms are coplanar by crystal symmetry; the TiO(1)NO(1') plane is perpendicular to the ring centroid-Ti-centroid plane; the two planes intersect at an angle of 89.9°; the plane of the pyridine ring is not quite coplanar with the TiO(1)NO(1) plane. A slight rotational deviation of these two planes by 3.6° is connected with a rotation of both CO_2 groups by 4.8° out of plane of the pyridine ring, and by 7.4° out of the TiO(1)NO(1') plane. A similar deviation from coplanarity, which was reported to have led to a slight shortening of the Ti-N relative to the two Ti-O distances, with respect to a fully coplanar geometry, has been reported for the Ti(IV) complex, [(H₂O)₂(O₂)Ti(dipic)].[63, 73]

Table 1. The bond lengths (pm) and bond angles (°) at the Ti(IV) metal centre in [(C$_5$H$_5$)$_2$Ti(dipic)]

Ti-O	211.1(6)	O-Ti-N	71.1(2)
Ti-N	216.0(8)	CR-Ti-CR'	133.0
Ti-CR	205.2	PL-PL'	47.2
Ti-PL	205.2		
Ti-C(1)	238(1)		
Ti-C(2)	237(1)		
Ti-C(3)	235(1)		
Ti-C(4)	238(1)		
Ti-C(5)	236(1)		

CR = centroid of C$_5$ ring, PL = mean plane of C$_5$ ring.

For the complex, [(C$_5$H$_5$)$_2$Ti(dipic)] (Figure 5), Leik et al.[64] concluded that it is apparent that the dipicolinate ligand, with its rather small bite angle of ~70°, is almost ideally suited to induce a pentacoordinate geometry even at a (C$_5$H$_5$)$_2$Ti centre, which otherwise appears to avoid this increase in coordination number, probably for steric reasons. Table 1 shows the bond lengths and bond angles at the Ti(IV) metal centre in [(C$_5$H$_5$)$_2$Ti(dipic)].

Vanadium, in different oxidation states, has been used in conjunction with dipicolinic acid and its analogues to produce coordination complexes.[62, 74-90] A selection of vanadium-containing complexes is discussed below.

The novel complex, C(NH$_2$)$_3$[VO$_2$(dipic)].2H$_2$O, and its analogous complex, NH$_4$[VO$_2$(dipic)], were synthesized and characterized.[77] Figure 6 shows ORTEP diagrams for both anions. The vanadium(V) metal center shows a similar pentacoordinated environment in the guanidinium and ammonium salts of [VO$_2$(dipic)]$^-$.[77] In the anion, the VO$_2^+$ group is coordinated to a dipic^{2-} acting as a tridentate ligand through its carboxylic oxygen atoms [V–O distances of 1.983(2) and 1.988(2) Å in C(NH$_2$)$_3$[VO$_2$(dipic)].2H$_2$O and 1.974(2) and 1.978(2) Å in NH$_4$[VO$_2$(dipic)]] and the nitrogen atom [V–N distances of 2.086(2) (for C(NH$_2$)$_3$[VO$_2$(dipic)].2H$_2$O) and 2.091(2) Å (for NH$_4$[VO$_2$(dipic)])]. The dipic^{2-} ligand is planar [rms deviation of atoms from the least-squares plane of 0.012 (for C(NH$_2$)$_3$[VO$_2$(dipic)].2H$_2$O) and 0.051 Å (for NH$_4$[VO$_2$(dipic)])] with the metal ion lying onto this plane in C(NH$_2$)$_3$[VO$_2$(dipic)].2H$_2$O [at 0.001(7) Å] and slightly above in NH$_4$[VO$_2$(dipic)] [at 0.136(1) Å]. The ligand plane of C(NH$_2$)$_3$[VO$_2$(dipic)].2H$_2$O bisects the dioxovanadium V=O double bonds whose lengths are 1.614(7) and 1.626(7) Å. This agrees with the structural data reported for the [VO$_2$(dipic)]⁻ complex in Cs[VO$_2$(dipic)].H$_2$O where the V=O double bond distances are 1.610(6) and 1.615(6) Å.[91] In contrast, the ligand plane of NH$_4$[VO$_2$(dipic)] structure departs appreciable from the O1=V=O2 bisector and the V–O1 bond distance is 0.012 Å (i.e. six times the rms error) longer than the V–O2 length [1.612(2) Å]. That difference in the VO distances is probably due to a pair of medium to strong N–H•••O1 bonds with the NH$_4$⁺ counter-ion (see below). There was a report of an example of even more pronounced V=O bond asymmetry in the VO$_2$⁺ group, namely in a bis oxo bridged binuclear vanadium(V) complex of stoichiometry [CH$_3$NHC(NH$_2$)$_2$]$_2$[V$_2$O$_4$(dipic)$_2$].[80] In this compound, the oxygen atom of the VO$_2$⁺ moiety, laying near the coordination plane, bridges the two halves of a centro-symmetric dimer through a weak axial V•••O bond. As a consequence of this, the two V=O distances differ in 0.078(1) Å [d(V–O1) = 1.606(1) Å].[77]

In C(NH$_2$)$_3$[VO$_2$(dipic)].2H$_2$O, the planar guanidinium counterion lays parallel to the dipic^{2-} ligand at a van der Waals contact distance of 3.1 Å. [77] The [VO$_2$(dipic)]⁻ and [C(NH$_2$)$_3$]⁺ ions are arranged in the lattice along layers parallel to (010) crystallographic planes. These layers are stabilized by a network of medium to strong intra-layer N–H•••O bonds involving the guanidinium NH$_2$ groups and carboxylic oxygen atoms of the dipic^{2-} ligand [N..O distances and N–H•••O angles are found in the ranges 2.848–2.982 Å and 132.6–176.1°, respectively]. Two parallel N–H•••O bonds are formed between the N2 and N3 atoms of the guanidinium ion and the carboxylate oxygens O3 and O4 of a neighboring anion. The interaction gives rise to an hexagonal pattern analogous to the observed in the bis-oxo-bridged V(V) complex[80] and similar to that found in a large number of H-bonded layered crystals and to that described for the interaction of phosphate, sulfonate, carboxylate and nitrate with this cation.[77] The N1 nitrogen atom of guanidinium interacts, through one of its hydrogen atoms, with the O6 oxygen from a carboxylate ion of another neighboring unit. The lattice is further stabilized by inter-layer H-bonds mediated by one crystallization water molecule held onto the layer by a N–H•••Ow bond [the N•••Ow distance and the N–H•••Ow angle are 2.825 Å and 157.5°, respectively]. This molecule is acting as a donor in two Ow–H•••O interactions with the dioxovanadium groups of neighboring layers [the corresponding Ow•••O distances and Ow–H•••O angles are 2.823 and 2.860 Å and 152.6° and 145.4°, respectively].

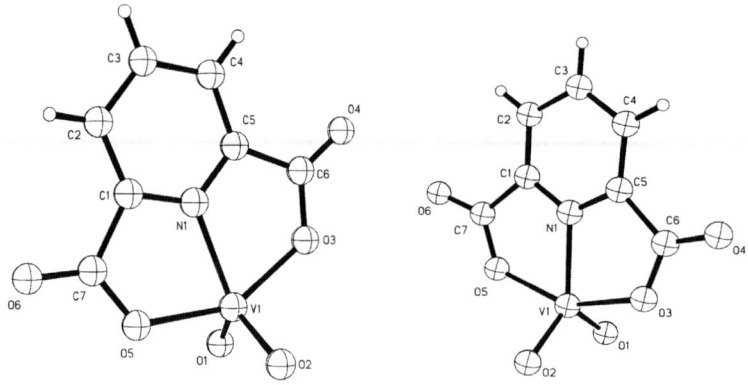

Figure 6. Diagrams of two [VO$_2$(dipic)]⁻ anions. (Reproduced by permission from reference 77).

[V^{IV}O(H$_2$O)$_2$(dipic)].2H$_2$O was synthesized by the reaction of VO(acac)$_2$ with dipicolinic acid.[85] The X-ray three-dimensional structural determination of [V^{IV}O(H$_2$O)$_2$(dipic)].2H$_2$O revealed that it crystallizes in the triclinic space group P_11 with two molecules in the unit cell, and consists of [V^{IV}O(H$_2$O)$_2$(dipic)] and two lattice water molecules. As illustrated in Figure 7, the geometry is a disordered octahedron with vanadium(IV) coordinated by two oxygen atoms from water, two carboxyl oxygen atoms (COO) and one nitrogen atom from a dipicolinate ligand and one terminal oxo[85]. The dipicolinate ligand (COO, COO, N) chelates one vanadium atom to form two five-membered rings. This was referred as a new example of a vanadium(IV) complex with dipicolinate, different from other related vanadium complexes, e.g., potassium oxodiperoxo(pyridine-2-carboxylate)vanadate(V), K$_2$[VO(O$_2$)$_2$(PA)].2H$_2$O;[92] potassium oxodiperoxo (3-hydroxypyridine-2-carboxylate)vanadate (V), K$_2$[VO(O$_2$)$_2$ (3HPA)]. 3H$_2$O;[92] K$_3$ [VO(O$_2$)$_2$ (2,4-pyridinedicarboxylate)].2H$_2$O;[93] bpV (2,4-pdc); K$_3$ [VO(O$_2$)$_2$ (3-acetatoxypicolinate)] .2H$_2$ O, bpV(3-acetpic);[93, 94] [VO(3HPA) (H$_2$O)]$_4$.9H$_2$O,[94] V(pic)$_3$. H$_2$O,[95] and [VO(6epa)$_2$ (H$_2$O)]. 4H$_2$O. [85] For [V^{IV}O (H$_2$O) $_2$(dipic)]. 2H$_2$O, the V–O(1) bond length [1.594 (3) Å] is shorter than those in K$_2$[VO(O$_2$)$_2$ (PA)]. 2H$_2$O, K$_2$[VO(O$_2$)$_2$ (3HPA)].3H$_2$O, bpV(2,4-pdc), bpV(3-acetpic), [VO(3HPA) (H$_2$O)]$_4$.9H$_2$O, V(pic)$_3$ .H$_2$O and [VO(6epa)$_2$ (H$_2$O)].4H$_2$O, while the V–N bond length [2.163(4) Å] is slightly longer than those in the corresponding complexes above. The V–O$_{carb}$ distances in [V^{IV}O(H$_2$O)$_2$(dipic)].2H$_2$O are shorter those of K$_2$[VO(O$_2$)$_2$(PA)]. 2H$_2$O, K$_2$[VO(O$_2$)$_2$(3HPA)].3H$_2$O, bpV(2,4-pdc) and pbV(3-acetpic), close to that of [VO(3HPA)(H$_2$O)]$_4$.9H$_2$O and longer than those in V(pic)$_3$.H$_2$O and [VO(6epa)$_2$(H$_2$O)].4H$_2$O. However, it is surprising that the V–O$_{water}$ distances [2.016(4), 2.059(4) Å] are much shorter than those in the corresponding complexes.

The O=V–N and O=V–O$_{carb}$ angles are 178.04(11)° and 106.43(16)° in [V^{IV}O(H$_2$O)$_2$(dipic)].2H$_2$O, respectively.[85] Both angles are different from those in related vanadium complexes, perhaps because coordinated nitrogen is trans to a terminal oxygen atom in [V^{IV}O(H$_2$O)$_2$(dipic)].2H$_2$O, while, for other corresponding vanadium complexes, the coordinated O (COO) atom is trans to the terminal oxygen atom. The N–V–O$_{carb}$ angle is close to those found in other vanadium complexes. Comparisons of the detailed bond distances and angles related to vanadium complexes are given in Tables 2 and 3, respectively.[85]

It is worth noting that C(8)–O(7) and C(8)–O(4), and C(7)–O(5) and C(7)–O(6) are shortened, indicating partial double bond character. C(8)–O(7) and C(8)–O(4) are 1.232(3) and 1.289(4) Å,[85] respectively; while C(7)–O(5) and C(7)–O(6) are 1.298(4) and 1.227(4) Å, respectively, indicating more electron delocalization in [V^{IV}O(H$_2$O)$_2$(dipic)].2H$_2$O than in V(pic)$_3$.H$_2$O.[95]

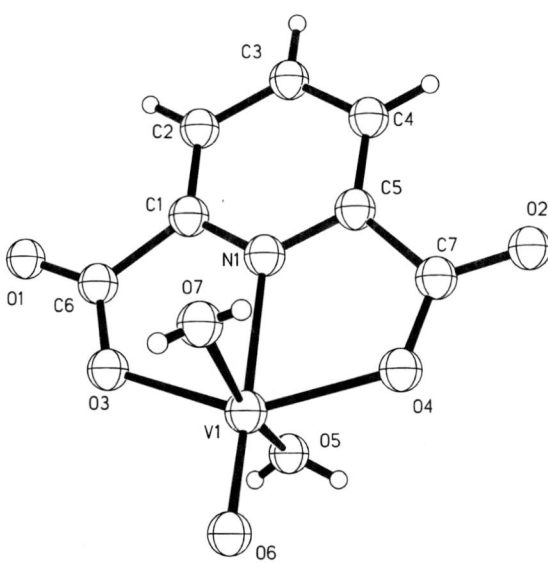

Figure 7. A diagram of [V^{IV}O(H$_2$O)$_2$(dipic)]. (Reproduced by permission from reference 85).

In addition, $[V^{IV}O(H_2O)_2(dipic)] \cdot 2H_2O$ contains some hydrogen bonds, primarily Oligand(H_2O)–H•••Ouncoordinated carbonyl, Oligand(H_2O)•••H–Ow and Ow–H•••Ocoordinated carbonyl.[85] The molecules are linked by two types of hydrogen bond. One is between coordinated oxygen atoms from the water and uncoordinated carbonyl oxygen atoms from dipicolinate ligands, the others are bridging hydrogen bonds formed between coordinated oxygen atoms from the water and coordinated carboxyl oxygen atoms from dipicolinate ligands by lattice water (Ow1) being bridged, namely Ooxo [O(2)]–H•••Ow(1)–H•••Ocoordinated carbonyl [O(4)]. Mononuclear vanadium $[V^{IV}O(H_2O)_2(dipicolinate)] \cdot 2H_2O$ units and crystallization water molecules are held together in an extensive two-dimensional network via O–H•••O hydrogen bonds, π–π stacking interactions between parallel aromatic pyridines, and face-to-face stacking interactions between parallel carboxylate groups of dipicolinate along the plane formed by the x, z axis of the unit cell. The V–V distance between molecules along the z axis is 6.568 Å ; along the x axis it is 9.129 Å.

Table 2. Comparison of the bond lengths (Å) in the related complexes

Complex	V=O	V-N	V-O_{carb}	V-O_{water}	Reference
$[V^{IV}O(H_2O)_2(dipic)] \cdot 2H_2O$	1.594(3)	2.163(4)	2.026(3)-2.051(4)	2.016(4)-2.059(4)	85
$K_2[VO(O_2)_2(PA)]$	1.599(4)	2.123(5)	2.290(4)		96
$K_2[VO(O_2)_2(3HPA)] \cdot 3H_2O$	1.606(2)	2.137(2)	2.314(2)		96
bpV(2,4-pdc)	1.622(9)	2.144(11)	2.299(8)		92
bpV(3-acetpic)	1.621(3)	2.179(4)	2.190(6)		92
$[VO(3HPA)(H_2O)]_4 \cdot 9H_2O$	1.584-1.608	2.124-2.152	1.963-2.154	2.034-2.073	93
$V(pic)_3 \cdot H_2O$		2.112(3)-2.153(3)	1.936(3)-1.966(2)		94
$[VO(6epa)_2(H_2O)] \cdot 4H_2O$	1.572(6)-1.596(6)	2.118(5)-2.153(5)	1.956(5)-2.002	2.219(5)-2.283(5)	95

Table 3. Comparison of the angle (°) in the related complexes

Complex	N-V=O	O=V-O$_{carb1}$	O=V-O$_{carb2}$	N-V-O$_{carb}$	Reference
[V^{IV}O(H$_2$O)$_2$(dipic)].2H$_2$O	178.07(11)	106.40(17)	82.27(16)	73.39(15)-73.75(15)	85
K$_2$[VO(O$_2$)$_2$(PA)	93.6(2)	166.7(2)		73.1(2)	96
K$_2$[VO(O$_2$)$_2$(3HPA)].3H$_2$O	94.92(7)	168.73(7)		73.0(6)	96
bpV(2,4-pdc)	93.1(4)	166.3(4)		73.7(6)	92
bpV(3-acetpic)	93.39(16)	166.04(15)		72.7(4)	92
[VO(3HPA)(H$_2$O)]$_4$.9H$_2$O	91.9-95.9	158.2-160.7	96.7-98.4	73.3-90.1	93
V(pic)$_3$.H$_2$O				76.48(10)-168.58(10)	94

[VO(dipic)(phen)].3H$_2$O was synthesized and characterized by X-ray crystallography.[83] The deprotonated dipicolinic acid acting as a tridentate chelating agent coordinates to the V(IV) metal centre through the heterocyclic ring nitrogen N(1) and the carboxylate oxygens O(3) and O(4). All three of them occupy three positions of a distorted square plane (Figure 8), the fourth position being occupied by one of the phen nitrogen atoms N(3).[83] The oxygen atom of the vanadyl moiety lie above the plane defined by O(3)-N(1)-O(4)-N(3), while the position *trans* to the vanadyl oxygen is occupied by the N(2) nitrogen atom of the coordinated phen ligand. The V-O(5) distance of 1.581 (3) Å is a little shorter than is generally found in most V(IV) complexes with a nitrogen donor attached to its trans position.[97, 98] Of the two V—N bonds generated by the coordinated phen ligand, the V--N(2) bond *trans* to the V=O bond is longer (2.312 Å) than the other V-N(3) bond (2.126 Å). The vanadium(IV) metal centre exists in a distorted octahedral donor environment. The deviation of the vanadium atom from the plane defined by O(3)-N(1)-O(4)--N(3) is 0.2910 Å and the dihedral angle between the mean planes defined by the aromatic ligands is 93.90°. Figure 1 shows that O(lw), O(2w) and O(3w) form part of the asymmetric unit. O(lw) is directly H-bonded with O(1) of the molecule. Each of the O(lw), O(2w) and O(3w) are connected to the symmetry generated O(lb), O(2a) and O(4c), respectively with transformation codes a(x,1 + y, z); b(1 - x, 1 - y, -z); c(x, 0.5 - y, - 0.5 - z) . Two of the oxygens O(lw) and O(2w) form three hydrogen bonds whereas O(3w) is connected by only two hydrogen bonds. Molecules are packed within the lattice through this type of hydrogen bonds. O(lw) exhibits two-fold disorder and accordingly atom O(lw) and O(1 'w) are assigned 0.7 and 0.3 occupancy, respectively.[83]

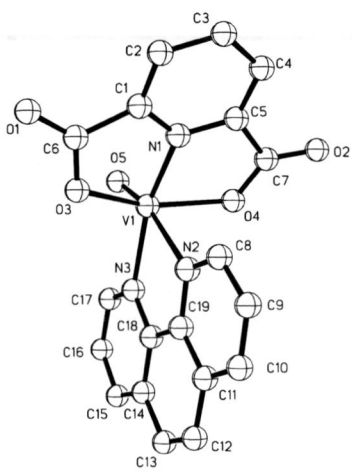

Figure 8. A diagram of [VO(dipic)(phen)]. (Reproduced by permission from reference 83).

4-Hydroxypyridine-2,6-dicarboxylatodioxovanadate(V) dihydrate was synthesized and characterized by X-ray crystallography.[99] (NMe$_4$) [VO$_2$(dipic-OH)].H$_2$O contains discrete [VO$_2$(dipic-OH)]$^-$ complex anions. The structure of the anion is shown in Figure 9. The asymmetric unit contains two formula units of (NMe$_4$)[VO$_2$(dipic-OH)].H$_2$O, which exhibit only minor structural differences.[99] The vanadium(V) metal center is five coordinate by virtue of coordination by two oxo ligands and the tridentate [dipic-OH]$^{2-}$ ligand (utilizing two carboxylate oxygen atoms and the pyridine nitrogen atom). The hydroxyl group (O(5), O(15)), one carboxylate oxygen atom (O(3), O(13)), and one oxo ligand (O(6), O(16)) from each of the [VO$_2$(dipic-OH)]$^-$ ions in the asymmetric unit form hydrogen bonds to water molecules, resulting in extended chains of [VO$_2$(dipic-OH)]$^-$ anions. The chains are separated by the tetramethylammonium cations.[99] The oxo ligands (O(6), O(16)) that are involved in hydrogen bonding with water form slightly longer bonds to vanadium (1.626(3), 1.627(3) Å) than do the oxo ligands (O(7), O(17)) that do not participate in hydrogen bonding (1.615(3), 1.612(3) Å). The shorter V=O bond lengths are similar to the V=O bond lengths observed in [VO$_2$(dipic)]$^-$ (1.610(6), 1.615(6) Å).[100] Hydrogen bonding to water does not seem to influence significantly the C-O(carboxylate) bond lengths. Other bond lengths and angles in the primary coordination sphere (V-O(carboxylate), V-N(pyridine), and V-O(oxo)) are similar to those observed for [VO$_2$(dipic)]$^-$.[100] An asymmetry in the V-O(carboxylate) bonding is observed; V-O(2) (1.998(4) Å) is slightly shorter than V-O(1) (2.022(3) Å), and a corresponding asymmetry is also observed for the other complex ion in the asymmetric unit.[99]

For Na[VO$_2$(dipic-OH)].2H$_2$O, the structure of the anion is shown in Figure 10. The asymmetric unit contains two formula units of Na[VO$_2$(dipic-OH)].2H$_2$O. As in (NMe$_4$)[VO$_2$(dipic-OH)].H$_2$O, the vanadium(V) atom is five coordinate by virtue of coordination to two oxo ligands and the tridentate [dipic-OH]$^{2-}$ ligand. The Na$^+$ cation is incorporated into a polymeric chain formed by coordination of Na$^+$ by [VO$_2$(dipic-OH)]$^-$ anions. The sodium ion is six-coordinate by virtue of coordination to two water molecules (O(8), O(9)), to two bridging oxo ligands (from two symmetry-related complexes (O(7), O(7B)), and to two carboxylate oxygen atoms from a third symmetry-related complex (O(2A), O(4A) in Figure 10). The carboxylate group at C(7) is therefore in a μ^3 coordination mode, and the group at C(6) is in a terminal monodentate coordination mode. The hydroxyl group O(5) (H-donor), the carboxylate group at C(6) (H-acceptor), and the oxo ligand O(6) (H-acceptor) form hydrogen bonds to the water molecules coordinated to the sodium ion, resulting in a linking together of the polymeric chains into extended sheets. Hydrogen bonding to water and coordination to the sodium ion influences bond lengths within the carboxylate groups.[99] For example, the difference in the C-O bond lengths in the carboxylate at C(6) (1.295(3), 1.233(3) Å), which engages in hydrogen bonding to water through O(3), is less pronounced than the difference in the C-O bond lengths seen in the carboxylate at C(7) (1.305(3), 1.223(3) Å), where both of the oxygen atoms coordinate the sodium ion. In contrast, for the oxo ligands hydrogen bonding to water (O(6)) or coordination to sodium (O(7)) does not result in an observable difference between the V=O distances (V=O(6), 1.626(2) Å; V=O(7), 1.629(2) Å). These distances are similar to the V=O distances in (NMe$_4$)[VO$_2$(dipic-OH)].H$_2$O (1.6264(17), 1.6290(17) Å) and slightly longer than those observed in the parent compound Cs[VO$_2$(dipic)].H$_2$O (1.610(6)/1.615(6) Å).[100] Otherwise bond distances and angles in the coordination sphere of the V(V) metal center in Na[VO$_2$(dipic-OH)].2H$_2$O are similar to the corresponding parameters in (NMe$_4$)[VO$_2$(dipic-OH)].H$_2$O and Cs[VO$_2$(dipic)].H$_2$O.[100] The most significant difference among these structures arises from the formation of a polymeric structure as a result of the interactions of the sodium ion with coordinated dipic-OH^{2-} ligands.[99]

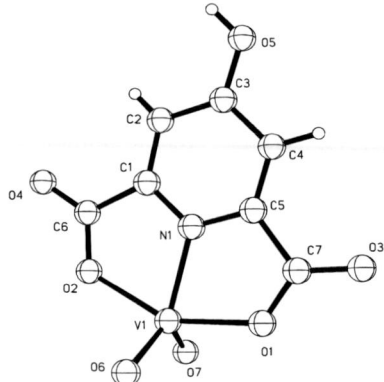

Figure 9. A diagram of the [VO₂(dipic-OH)]⁻ anion. (Reproduced by permission from reference 99).

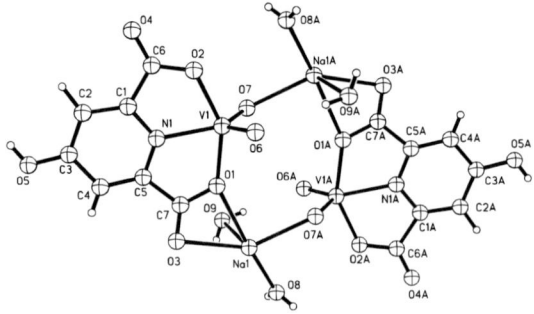

Figure 10. A diagram of the [VO₂(dipic-OH)]⁻ anion. (Reproduced by permission from reference 99).

The reaction between [VO(dipic)(H$_2$O)$_2$].H$_2$O and creatinine resulted in the formation of a bis(oxo-bridged) binuclear vanadium(V) compound of stoichiometry [CH$_3$NHC(NH$_2$)$_2$]$_2$[V$_2$O$_4$(dipic)$_2$] (where [CH$_3$NHC(NH$_2$)$_2$]$^+$ = methyl gaunidinium).[80] An ORTEP drawing of the binuclear vanadium(V) complex with the atom numbering scheme is shown in Figure 11. The [VO$_2$(dipic)]⁻ ions are arranged in the lattice as centrosymmetric oxo-bridged binuclear complexes. The pair of vanadium(V) atoms in a dimer is in an edge sharing octahedral environment, with the dioxo vanadium(V) cation coordinated to a dipicolinate molecule acting as a tridentate ligand through one oxygen of each carboxylic group [V–O distances of 1.984(1) and 1.995(1) Å] and the heterocyclic nitrogen atom [d(V–N) = 2.097(2) Å].[80] The dipicolinate group defines an equatorial ligand plane [with a rms deviation of atoms from the least-squares plane of 0.025 Å] with the metal lying close to this plane [at 0.144(1) Å]. The bridging oxo ligand [d(V–O2) = 1.684(1) Å] is much closer to this plane [at 0.486(2) Å] than the terminal oxo atom [at 1.739(2) Å] which shows a slightly shorter V=O bond distance [d(V–O1) = 1.606(1) Å] and occupies an axial position. The O1–V–O2 angle is 105.36(7)°. The octahedral bonding structure around the metal is completed at the other axial position by the weak interaction with the bridging oxo ligand of the inversion related VO_2^+ group in the dimer [d(V–O2') = 2.370(1) Å].[80] The bonding structure around vanadium(V) agrees well with other binuclear dioxovanadium(V) complexes of tridentate ligands reported in the literature.[101] The above structural data was compared with that reported for the monomeric complex, $Cs[VO_2(dipic)]\cdot H_2O$.[100] Here the V=O double bond distances are equal to within experimental accuracy [1.610(6) and 1.615(6) Å], while in the dimeric complex, the V–O2 bond involving the bridging oxygen atom is 0.078(2) Å longer than the terminal V–O1 distance. This slight bond asymmetry within the dioxovanadium(V) ion is due to the formation of the weak intermolecular V•••O bond bridging the halves of the binuclear complex. Significant bond localization is also observed in the carboxylic groups of the dipicolinate ligand. In fact, the terminal C–O bond lengths are about 0.1 Å shorter than the C–O distances involving the coordinated-to-vanadium oxygen atoms. The C–N geometrical parameters of the methylguanidinium cation (MG) are in good accord with those found in related structures.[102-104] It has a singular planar CN_3 skeleton with a strong electronic delocalization that makes it able to participate in nets of H-bonding.

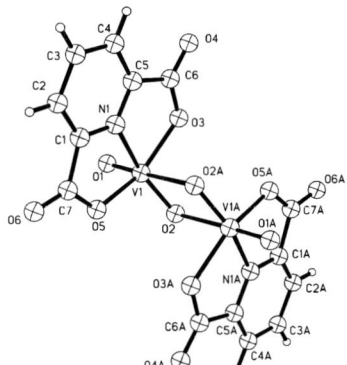

Figure 11. A diagram of the $[V_2O_4(dipic)_2]^{2-}$ anion. (Reproduced by permission from reference 80).

The [VO$_2$(dipic)]⁻ monomeric units and the [CH$_3$NHC(NH$_2$)$_2$]⁺ ions are arranged in the lattice along layers parallel to (101) crystallographic planes. Adjacent layers in the crystal are linked by the dimer-bridging bond. These layers are stabilized by a net of medium to strong N–H•••O bonds involving the ethylguanidinium NH and NH$_2$ groups and the oxo-bridge and the carboxylic oxygen atoms of the dipicolinate ligand [N•••O distances are in the range 2.858–2.964 Å and N–H•••O angles from 156.0° to 176.6°].[80] Each methylguanidinium has two N–H groups, from different N atoms, linked to one monomeric [VO2(dipic)]⁻ unit through an oxo-bridge atom and the carboxylate oxygen coordinated to the metal. The structural feature of this pattern is similar to that described for the interaction of phosphate, sulfonate and nitrate with the guanidinium cation.[102-104] The other N–H groups of the same MG ion bond to adjacent monomer units through N–H•••O hydrogen bonds, forming a layer structure.

A number of 4-substituted, dipicolinatodioxovanadium(V) complexes and their hydroxylamido derivatives were synthesized and characterized by X-ray crystallography.[105] The Na[VO$_2$(dipic-NH$_2$)].2H$_2$O complex (shown in Figure 12) is reported to crystallize as a salt without any required crystallographic symmetry. Selected bond distances and angles are provided in Table 4. The bond angles about the vanadium(V) suggested either a distorted square-pyramidal or trigonal-bipyramidal structure with the latter geometry being emphasized in Figure 12. The two largest angles about the V atom are O3-V1-O4) 149.99° and O2-V1-N1 = 126.05°. It was reported that when these angles were used to define τ,[106] then one of the oxo ligands (O1) becomes the 'apical' ligand and τ = (149.99 - 126.05)/60) 0.399. Since τ is close to 0.5, the coordination geometry for this complex is neither trigonal-bipyramidal nor square-pyramidal.[105] A more detailed comparison of this structure to similar structures is shown in Table 2 and discussed below.[81, 99, 100]

In Na[VO$_2$(dipic-NH$_2$)].2H$_2$O, the six-coordinate sodium cation is bound to two symmetry-related O2 oxo ligand atoms (2.418(3), 2.459(3) Å). In addition, symmetry-related oxygen atoms from the C1 carboxylate group (Na-O3) 2.406(3) Å, Na-O5) 2.627(3) Å) and both of the lattice water molecules (Na-O7 = 2.405(4) Å, Na-O8 = 2.271(4) Å) are bound to sodium. A long and very weak seventh interaction is also present between sodium and the other oxo oxygen atom (Na-O1 = 3.009(3) Å). The amino substituent forms a weak hydrogen bond to a water molecule (N2•••O7 = 2.884(5) Å), as does one of the carboxylate oxygen atoms (O6•••O7) 2.774(4) Å).

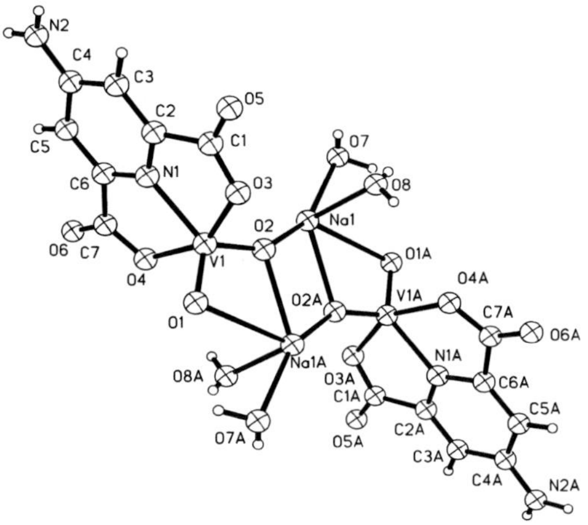

Figure 12. A diagram of Na[VO$_2$(dipic-NH$_2$)].2H$_2$O. (Reproduced by permission from reference 105).

The K[VO$_2$(dipic-NO$_2$)] complex (shown in Figure 13) also crystallizes without any required crystallographic symmetry; the pertinent bond angles and distances are listed in Table 3. The bond angles about the vanadium(V) suggested either a distorted square-pyramidal or trigonal-bipyramidal structure shown in Figure 13. The two largest angles about the V atom are O3-V-O4 = 148.81° and O1-V-N1 = 131.85°. Since the O2 atom is not involved in either of these two angles, it becomes the 'apical' ligand, and τ[106] = (148.81 - 131.85)/60) 0.283. Since the τ value is closer to 0 than to 1, the coordination geometry for this complex is approaching square-pyramidal.[105] The bond distances and angles in these structures may be compared to the corresponding structural features of the Cs[VO$_2$(dipic)][100] and Na[VO$_2$(dipic-OH)],[81, 99] and Na[VO$_2$(dipic-NH$_2$)] (see Table 4). From the values in this table it is clear that the V=O and V-O bond lengths are virtually identical, showing that substitution in the 4-position of the pyridine ring does not affect those bonds. While a majority of the metric parameters in K[VO$_2$(dipic-NO$_2$)] are nearly identical to that of the amino and hydroxyl-substituted dipic^{2-} complexes, there are two notable differences. The first is the V-N$_{py}$ distance of 2.1019(17) Å, which is significantly longer than those bond lengths in complexes with electron donating substituents (Table 4). The other is the fact that the pyridine ring has a different orientation than in the other substituted dipicolinate complexes.[105] The O(1) atom is oriented in a slightly more trans fashion to the pyridine nitrogen (<N-V-O(1) = 131.85(7)°) than the O(2) atom which is

in more of a *cis* orientation (<N-V-O(2) = 118.59(7)°).[105] This structural modification suggests that, with the appropriate ligand, it may be possible to introduce an additional ligand to form a six-coordinate complex. The N-V-O angles in the other substituted dipicolinate complexes are far more symmetric with respect to the dipicolinate ring, while the unsubstituted dipic structure displays somewhat more asymmetry (Table 4). In addition, the plane defined by the NO_2 group is twisted ~27° relative to the pyridine ring.

The V-N_{py} bond varies in length as would be expected from the electronic changes in the substituent group. The more electron donating substituent, NH_2, should make the pyridine N-atom a better σ and π donor and shorten the V-N_{py} bond, while the electron-withdrawing NO_2 group should have the opposite effect. Indeed the V-N_{py} bond length in [VO_2(dipic-NH_2)]⁻ is 2.050(3) Å compared to the V-N_{py} bond lengths in [VO_2(dipic-OH)]⁻ (2.077(4) and 2.0770(19) Å)24 and in [VO_2(dipic)]⁻ (2.089(6) Å);[100] the V-N bond length in [VO_2(dipic-NO_2)]⁻ is noticeably longer at 2.1019(17) Å. The eight-coordinate potassium ions knit the structure together tightly by binding to oxygen atoms from seven different symmetry-related complex anions. Five of the K-O bonds are markedly shorter than the other three. Three of these five shortest bonds to potassium involve oxo ligands O1 (2.7395(15) Å) and O2 (2.7495(15), 2.7681(15) Å). The other two short K-O interactions involve oxygen atoms from the nitro substituent (O7, 2.7139(15) Å) and one of the carboxylate groups (O5, 2.7838(16) Å). There is no hydrogen bonding in this structure due to a lack of protic donors.

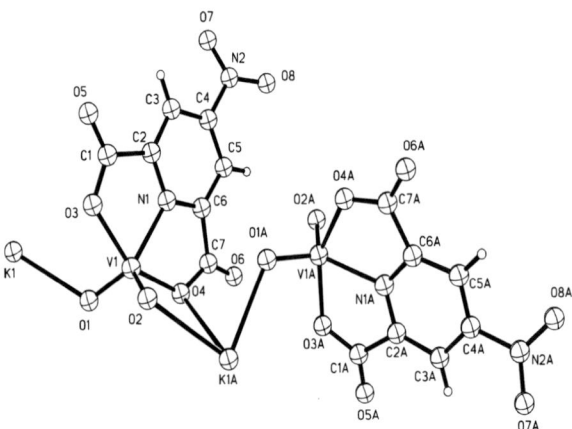

Figure 13. A diagram of K[VO_2(dipic-NO_2)]. (Reproduced by permission from reference 105).

Table 4. Selected bond lengths and angles for Na[VO$_2$(dipic-NH$_2$)]·2H$_2$O, K[VO$_2$(dipic-NO$_2$)], and other five coordinate dipicolinato-vanadium(V) complexes

Complex	V=O	V-N$_{py}$	V-O$_{carb}$	O$_{carb}$-V-O$_{carb}$	O$_{xo}$-V-N$_{py}$	O$_{carb}$-V-N$_{py}$	References
Na[VO$_2$(dipic-NH$_2$)]·2H$_2$O	1.620(3) 1.627(3)	2.050(3)	1.990(3) 1.991(3)	149.99(12)	124.92(13) 126.05(13)	75.15(11) 74.85(11)	105
K[VO$_2$(dipic-NO$_2$)]	1.6253(15) 1.6293(14)	2.1019(17)	1.9910(14) 1.9953(15)	148.81(6)	118.59(7) 131.85(7)	74.75(6) 74.62(6)	105
Cs[VO$_2$(dipic)]·H$_2$O	1.610(6) 1.615(6)	2.089(6)	2.001(5) 1.982(5)	149.4(2)	122.0(3) 128.2(3)	74.6(2) 75.9(2)	100
NMe$_4$[VO$_2$(dipic-OH)]·H$_2$O	1.615(3) 1.626(3)	2.077(4)	1.008(4) 2.022(3)	148.88(14)	123.37(17) 125.92(18)	74.41(14) 74.47(7)	99
Na[VO$_2$(dipic-OH)]·2H$_2$O	1.6264(17) 1.6290(17)	2.0770(19)	1.9945(16) 2.0011(16)	149.42(7)	124.48(8) 125.71(8)	74.96(7) 74.47(7)	99
K[VO$_2$(dipic-OH)]·H$_2$O	1.606(5) 1.616(5)	2.089(6)	2.03395 1.990(5)	148.0(2)	123.1(3) 127.4(3)	73.3(2) 74.7(2)	81

The [VO(dipic)(Me$_2$-NO)(H$_2$O)].0.5H$_2$O complex (shown in Figure 14) crystallizes as discrete molecules without required crystallographic symmetry. Selected pertinent bond distances and angles are provided in Table 5 along with those of other crystallographically characterized, seven-coordinate vanadium(V)-dipic complexes.[89, 107-109] This complex contains seven coordinate vanadium in a pseudo, pentagonal-bipyramidal geometry. Alternatively, if one considers the hydroxylamido group to be a monodentate ligand, then the complex is a distorted, six-coordinate octahedral complex. The geometry of this complex is similar to that of the vanadium(V)-dipicolinato complex reported previously with the correspondingparent hydroxylamine, [VO(dipic)(H$_2$NO)(H$_2$O)].[89] Dimethylation of the hydroxylamine unit has no observable effects on the lengths of the V-O$_{H2O}$, V-O$_{R2NO}$, V-Npy, and the hydroxylamine N-O bonds. However, in the substituted hydroxylamine complex, the V-N$_{R2NO}$ bond length increased from 2.007(3) to 2.028(3) Å, respectively; such an increase would be anticipated based on the extra steric bulk of the two methyl groups. This steric bulk is also apparently sufficient to modify the coordination sphere of the vanadium resulting in a decrease of the V-O$_{carb}$ bond lengths, from 2.031(3) and 2.039(3) Å in 1a to 2.008(3) and 2.026(3) Å in the substituted complex, 1c.[105] Comparing the bond lengths of the hydroxylamido derivative complexes with those of the parent dipicolinato complexes (Tables 4 and 5) showed several differences. The V=O bonds are significantly shorter in the hydroxylamido complexes perhaps reflecting the need for additional electronic density in the more sterically crowded complexes. The V-N$_{py}$ bonds are shorter, whereas the V-O$_{carb}$ bonds are significantly longer in the hydroxylamido complexes compared to the parent complexes. Due to the fact that there is little change in the N-C-C and C-C-O angles, the shortening of the V-O$_{carb}$ bonds is primarily attributable to better overlap between the vanadium atom and the pyridine nitrogen donor.

Hydrogen bonding connects the complexes via interactions between the coordinated water molecule (O6) and carbonyl oxygen atoms O3 (2.817(5) Å) and O4 (2.764(5) Å). The disordered water molecule present in the lattice does not form any significant hydrogen bonds (perhaps accounting for its disorder).[105]

Coordination Chemistry of Dipicolinic Acid and Its Analogues 23

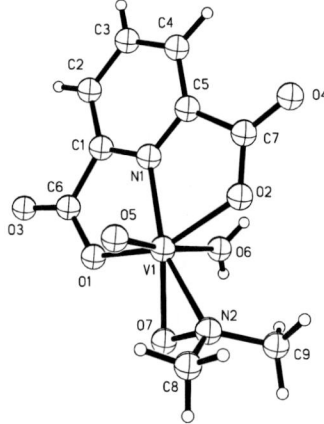

Figure 14. A diagram of [VO(dipic)(Me$_2$NO)(H$_2$O)]. (Reproduced by permission from reference 105).

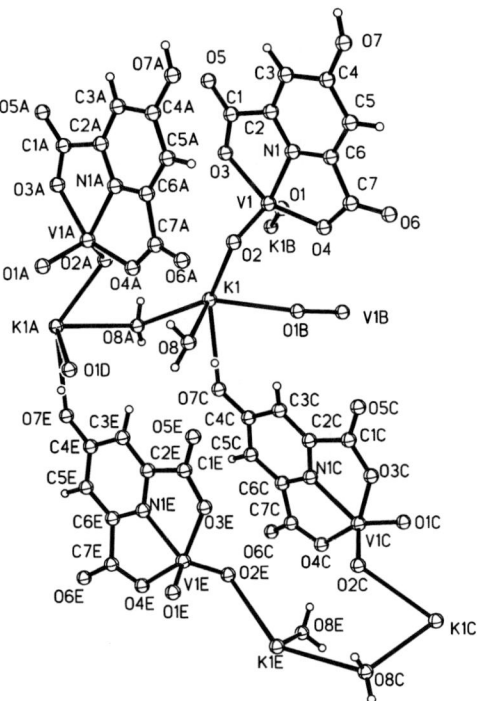

Figure 15. A diagram of K[VO$_2$(dipic-OH)].H$_2$O. (Reproduced by permission from reference 81).

Table 5. Selected bond length and angles for [VO(dipic)(Me$_2$NO)(H$_2$O)]·0.5H$_2$O and other known seven-coordinate dipicolinato-vanadium(V) complexes with ternary peroxo or hydroxylamido ligands

	[VO(dipic)(O$_2$)(H$_2$O)]	[VO(dipic)(O$_2$)(H$_2$O)]	[VO(dipic)(OOBu^t)(H$_2$O)]	[VO(dipic)(H$_2$NO)(H$_2$O)]	[VO(dipic)(Me$_2$NO)(H$_2$O)]·0.5H$_2$O
Reference	107	109	108	89	105
V-O$_{oxo}$	1.579(2)	1.588(3)	1.574(3)	1.587(3)	1.588(3)
V-N$_{py}$	2.088(2)	2.080(3)	2.058(4)	2.064(3)	2.067(3)
V-N$_{R2NO}$				2.007(3)	2.028(3)
V-O$_{carb}$	2.053(2) 2.064(2)	2.041(3) 2.048(3)	1.996(3) 1.983(3)	2.031(3) 2.039(3)	2.026(3) 2.008(3)
V-O$_{R2NO}$				1.903(3)	1.902(3)
V-O$_{peroxo}$	1.870(2) 1.872(2)	1.869(3) 1.892(3)	1.872(3) 1.999(3)		
V-O$_{H2O}$	2.211(2)	2.235(3)	2.234(3)	2.240(3)	2.239(4)
N$_{R2NO}$-O$_{R2NO}$				1.371(4)	1.384(4)
O$_{peroxo}$-O$_{peroxo}$	1.441(3)	1.437(4)	1.436(5)		
O$_{carb}$-V-O$_{carb}$	147.2(1)	148.20(12)	149.4^a	149.5(1)	148.44(12)
O$_{oxo}$-V-N$_{py}$	92.2(1)	95.53(16)	95.9(1)	97.2(1)	93.68(14)
O$_{oxo}$-V-O$_{carb}$	96.2(1) 94.9(1)	95.98(16) 94.38(16)	95.2(1) 94.1(1)	93.0(1) 98.9(1)	98.05(14) 95.08(13)
O$_{H2O}$-V-O$_{oxo}$	172.1(1)	169.84(16)	172.8(1)	172.0(1)	172.12(14)

^{a} Not reported; calculated from the cif file.[105]

K[VO$_2$(dipic-OH)]·H$_2$O was synthesized and characterized by X-ray crystallography and other physical techniques.[81] In Figure 15, the structure and the atom labeling scheme for K[VO$_2$(dipic-OH)]·H$_2$O is shown. [VO$_2$(dipic-OH)]$^-$ exists as a discrete mononuclear unit with the vanadium atom in a distorted trigonal bipyramidal coordination environment. The pyridine nitrogen atom (N1) and two oxygen atoms (O1, O2) of the VO$_2$ group coordinate to the vanadium center and occupy the distorted equatorial plane, while two carboxylate oxygen atoms occupy the axial positions. The vanadium atom is positioned 0.022(4) Å above the least squares equatorial plane through O1, O2, and N1. The chelation of the ligand decreases the angles around the vanadium center with carboxylate oxygen atoms and pyridine nitrogen atom (N1-V-O3 = 73.3(2)° and N1-V-O4 = 74.7(2)°) of the dipic-OH ligand with concomitant increase in other two angles (O3-V-O2 = 98.8(3)8 and O2-V-O4 = 100.4(3)°).[81] The constraints imposed by the chelation of the carboxylate oxygen atoms of the dipic-OH ligand at trans-positions result in a decrease of the O3-V-O4 bond angle from 180.08 to 148.0(2)°. As a result of the coordination of the carboxylate oxygen atoms of the dipic-OH ligand, the C1-C2-N1 and N1-C6-C7 angles on the pyridine ring are also decreased from 120.08 to 108.8(6)° and 110.8(6)°, respectively, with an increase of other corresponding bond angles.[81]

The VO$_2$ group is in the cis configuration, with an O1-V1-O2 angle of 109.5(3)° and with V-O1 and V-O2 distances of 1.606(5) and 1.616(5) Å, respectively. These V=O bonds are sufficiently short to imply double bonding with considerable π character. The V=O bond distance of [VO$_2$(dipic-OH)]$^-$ (1.606(5)/1.616(5) Å) is similar to the reported values for five-coordinate monooxovanadium(IV) complexes, including the Na$^+$ and NMe$_4^+$ salts of [VO$_2$(dipic-OH)]$^-$ (1.6264(17)/1.6290(17) and 1.615(3)/1.626(3) Å),[84] [VO(bzac)$_2$] (1.612 (10) Å),[110] [VO(acac-Et)$_2$] (1.605(2) Å),[111] and [VO(acac-Me)$_2$] (1.592(2) Å).[111] V=O bond distances of the VO$_2$ group in six-coordinate vanadium(V) complexes is larger than that observed in this complex including [VO$_2$(EDDA)]$^-$ (1.632(1)/1.655(2) Å),[112] [VO$_2$(EDTA)]$^-$ (1.639(2)/1.657(1) Å),[113] and [VO$_2$(pic)$_2$]$^-$ (1.637(2)/1.638(2) Å).[114] The similarity of the V=O bond distance in five-coordinate V(IV) and V(V) complexes show that this bond distance is dependent upon the geometry around vanadium center and not on oxidation state. Long bonds extend from the vanadium atom to the carboxylate oxygen atoms (V-O3 = 2.033(5) and V-O4 = 1.990(5) Å) coordinated trans to each other at axial positions. These bond distances are in the range reported for other complexes including [VO$_2$(pic)$_2$]$^-$ (1.989(2) Å)[111] in which the carboxylate oxygen atoms are coordinated at axial positions. The nitrogen atom is coordinated in the distorted trigonal plane with V-N bond distance of 2.089(6) Å. This bond is, as expected, shorter than the bond distances reported for square pyramidal complexes in which nitrogen is coordinated in axial and/or equatorial plane[114, 115] and similar to the corresponding Na^+ and NMe_4^+ salts (2.0770(19) and 2.077(4)/2.070(4) Å).[84] The unit cell also contains one water molecule and a potassium ion. The water molecule resides between two monomer units making a hydrogen bond with one carboxylate oxygen atom of each of two monomer units (O5•••O8 = 2.893 Å, O4•••O8 = 2.921 Å) making a discrete dimer. The hydrogen bonding in this complex may be considered weak,[116] but the O8-H8b-O5 angle of 137.28 indicates that the hydrogen bonding is similar to those observed in organic structures.[117] The distances of the potassium ion from the oxygen atoms of the dioxo group (O1, O2), the oxygen atom of phenol OH (O7) and the oxygen atom of water molecule (O8) ranged from 2.664 to 2.878 Å which is normal for K^+•••O contacts.[118] However, the interactions of the potassium ion with O1 and O2 are not strong (2.793 and 2.878 Å) enough to have a significant affect on the V=O bond distance.[81]

$K[V(C_8H_3NO_6)O_2]\cdot H_2O$, was synthesized by reacting 4-carboxypyridine-2,6-dicarboxylic acid (contaminated as a potassium salt) with NH_4VO_3 in aqueous solution.[90] The complex, with a vanadium(V) metal center, is a distorted square-based pyramid (Figure 16). Its structure consists of chains of the anionic complexes in the direction of the *b* axis connected by potassium–oxygen interactions which range from 2.5981(18) to 3.0909(18) Å.[90] These chains are linked to each other by hydrogen bonding between the O atoms of the complex and the water molecules.[90] Selected bond lengths for $K[V(C_8H_3NO_6)O_2]\cdot H_2O$ are shown in Table 6.

Table 6. Selected bond lengths for $K[V(C_8H_3NO_6)O_2]\cdot H_2O$

Bond	Length/Å
V1-O6	1.6187(17)
V1-O5	1.6287(17)
V1-O2	1.9949(17)
V1-O1	2.0091(17)
V1-N1	2.086(2)

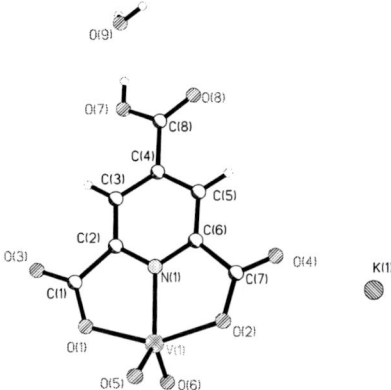

Figure 16. A diagram of K[V(C$_8$H$_3$NO$_6$)O$_2$].H$_2$O. (Reproduced by permission from reference 90).

Payne *et al.* [119] recently reported the crystal structure of the [Cr(dipic)$_2$]$^-$ anion, with protonated 2,2'-dipyridylamine (Hdpa) as a counter ion. [(2-pyridyl)(1-hydro-2-pyridinium)amine][bis(2,6- pyridine dicar boxyla to) chromate (III)] trihydrate, **1**, shows N$_2$O$_4$ coordination of the chromium (III) anion that is provided by two dianionic ligands, dipicolinate (Figure 17).[119] The distorted octahedral geometry of the chromium(III) metal center compares favorably in bond lengths and angles to that of the previously reported structure containing a rubidium cation.[120] Table 7 shows selected bond distances and angles for [Hdpa][Cr(dipic)$_2$]·3H$_2$O.

The [Hdpa]$^+$ cation (Figure 18) shows an isolated protonation at the N4 atoms with no positive residual electron density located near N5 in the final difference Fourier maps. No differences were observed in the bond distances of either of the formally pyridine and pyridinium rings. Bond localization is observed in the pyridinium-amine bond. A shortening of 0.037 Å is observed in the pyridinium-amine bond length while the pyridine-amine bond length compares favorably to a previously published structure of dpa.[121] In that structure, the dpa molecule crystallizes as a hydrogen bonded dimer in which the pyridine-amine bond length was found to be 1.380(4) Å. The rings of the cation in [(2-pyridyl) (1-hydro-2-pyridinium) amine] [bis (2, 6-pyridine dicarboxyla to)chromate(III)] trihydrate are twisted out of plane by 5.66(1)°. The rings of the [Hdpa]$^+$cations stack along the *a* axis. A packing diagram (Fig 19) of [(2-pyridyl) (1-hydro-2-pyridinium) amine] [bis (2, 6-pyridinedicarboxylato) chromate(III)] trihydrate is viewed approximately down the *a* axis. The water molecules O11 and O12 alternate to form approximately tilted square hydrogen bonded tetramers at the corners of the b/c cell edge along the a axis. These squares are then hydrogen bonded to the non-ligated carboxylate oxygen atoms O2 and O8 of four alternating anions. The remaining water molecule has O10 sitting at the center of a hydrogen bonded triangle formed with NH_3 of a cation and two oxygen atoms O5 (ligated) and O2 (non-ligated) of alternating anions.[119] The hydrogen bonding between O5−O10−O2 form a network between the anion layers.[119] The packing also consists of three π-π ring interactions of less than 3.8 Å.[122] These interactions are detailed in Table 7. There are two anion-anion interactions and one cation-cation interaction which are all related by an inversion center of each of the individual ring components.

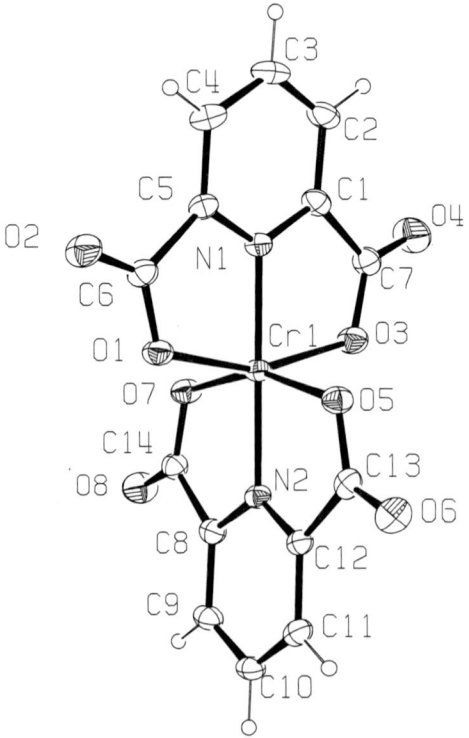

Figure 17. An ORTEP view of the anion of [(2-pyridyl)(1-hydro-2-pyridinium) amine][bis (2,6-pyridinedicarboxylato)chromate(III)] trihydrate shown with 30 % probability ellipsoids and the atom numbering scheme. (Reproduced by permission from reference 119).

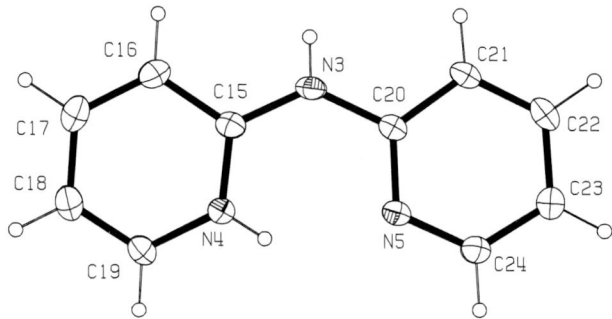

Figure 18. An ORTEP view of the cation of [(2-pyridyl)(1-hydro-2-pyridinium)amine][bis(2,6-pyridinedicarboxylato)chromate(III)] trihydrate shown with 25 % probability ellipsoids and the atom numbering scheme. (Reproduced by permission from reference 119).

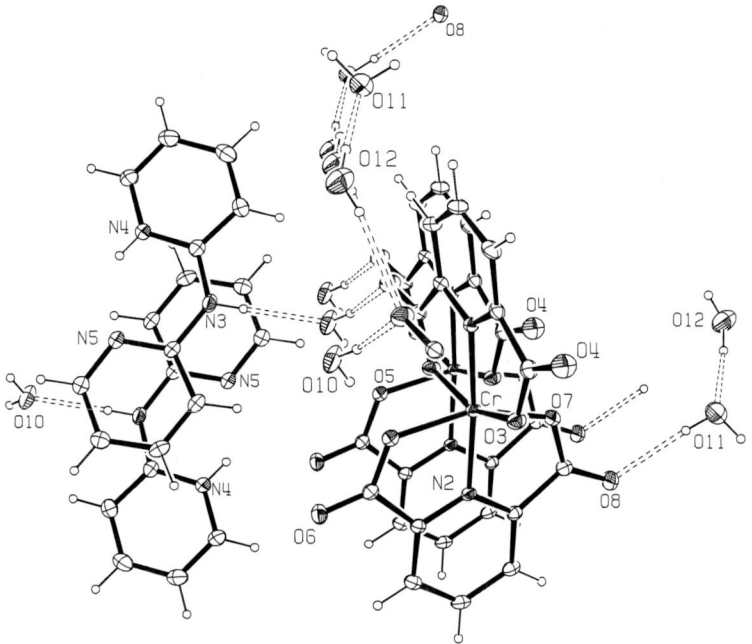

Figure 19. Selectively labeled ORTEP packing diagram of [(2-pyridyl)(1-hydro-2-pyridinium) amine] [bis(2,6-pyridinedicarboxylato)chromate(III)] trihydrate viewed approximately down the *a* axis. The thermal ellipsoids are drawn at the 20% probability level. (Reproduced by permission from reference 119).

Table 7. Selected Bond Distances (Å) and Angles (°) for [Hdpa][Cr(dipic)$_2$]·3H$_2$O

Cr-O3	1.9683(17)	O3-Cr-N1	79.30(7)	N2-Cr-O1	98.74(7)
Cr-N1	1.9706(19)	O3-Cr-N2	103.31(7)	O7-Cr-O1	95.77(7)
Cr-N2	1.9770(19)	N1-Cr-N2	175.35(8)	O3-Cr-O5	93.72(7)
Cr-O7	1.9834(17)	O3-Cr-O7	90.33(7)	N1-Cr-O5	97.79(7)
Cr-O1	1.9995(17)	N1-Cr-O7	105.21(7)	N2-Cr-O5	78.29(7)
Cr-O5	2.0071(16)	N2-Cr-O7	78.75(7)	O7-Cr-O5	157.00(7)
N3-C15	1.354(3)	O3-Cr-O1	157.88(7)	O1-Cr-O5	88.93(7)
N3-C20	1.391(3)	N1-Cr-O1	78.59(7)	C15-N3-C20	129.4(2)

The crystal structures of Na[Cr(dipic)$_2$].2H$_2$O and [Cr (dipic) (phen) Cl].0.5H$_2$O (see figure 20) were reported.[123] In Na[Cr(dipic)$_2$].2H$_2$O, the Cr(III) metal center is in a distorted octahedral environment, coordinated to two dipic^{2-} anions acting as tridentate ligands through its carboxylic oxygen atoms [Cr–O distances in the range from 1.985(5) to 1.998(4) Å] and the nitrogen atoms [d(Cr–N1) = 1.972(5) and d(Cr–N2) = 1.980(5) Å]. The dipic^{2-} ligands are nearly planar [rms deviation of atoms from the corresponding least squares planes less than of 0.07 Å] and perpendicular to each other [dihedral angle of 81.73(5)°]. The metal ion lies close onto the intersection of the coordination planes (along the N1•••N2 direction).[123]

The sodium ion is in a compact six fold coordination with four carboxylic oxygen atoms of neighboring dipic^{2-} groups [Na•••O distances in the range from 2.406(6) to 2.651(6) Å] and the two water molecules [d(Na•••O1w) = 2.329(6) Å and d(Na•••O2w) = 2.412(5)Å].[123]

For [Cr(dipic)(phen)Cl].0.5H$_2$O, the Cr(III) metal center is in a six-fold environment, coordinated to one dipic^{2-} ion defining a ligand equatorial plane [Cr–O distances of 1.975(3) and 1.999(3) Å and d(Cr–N) = 1.969(4) Å] and to a phen molecule acting as a bidentate ligand that bridges the fourth equatorial coordination site and one axial position through its nitrogen atoms [Cr–N distances of 2.075(4) and 2.081(3) Å, respectively].[123] The other axial position is occupied by a chlorine ion [d(Cr–Cl) = 2.290(1) Å]. The dipic^{2-} and phen species are nearly planar [rms deviation of atoms from the corresponding least-squares planes less than of 0.045 Å for both ligands] and close to mutual perpendicularity [angled at 86.60(7)° from each other]. As for the sodium salt, the Cr(III) ion nearly lies on the intersection of the coordination planes (along the N1•••N21 direction).[123]

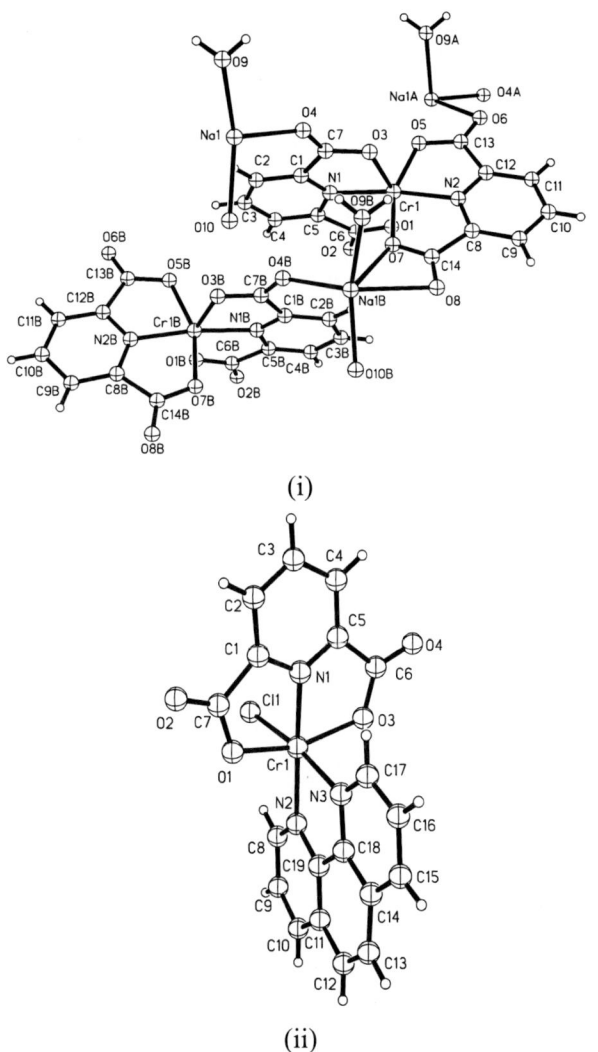

Figure 20. Diagrams of Na[Cr(dipic)$_2$]·2H$_2$O (i) and [Cr(dipic)(phen)Cl] (ii). (Reproduced by permission from reference 123).

The X-ray crystal structures of [Mn(dipic)(bpy)$_2$]·4.5H$_2$O and [Mn(chedam)(bpy)]·H$_2$O (chedam = chelidamic acid (4-hydroxypyridine-2,6-dicarboxylic acid) and bpy = 2,2'-bipyridine) were reported by Devereux *et al.*[124] The X-ray crystal structure of [Mn(dipic)(bpy)$_2$]·4.5H$_2$O is shown in Figure 21. The asymmetric unit consists of one tridentate dipicolinate ligand, two bidentate 2,2'-bipyridine molecules, four full occupancy water molecules and one half-occupancy water molecule. The [Mn(dipic)(bpy)$_2$] complex lies on a twofold axis, passing through Mn, N3 and C14 so that the by groups are *cis* to one another and the metal ion has irregular six-coordinate geometry. All the solvate water molecules and all the carboxylate oxygens of the dipic ligand are involved in hydrogen bonding. The ligands are involved in π-π stacking interactions with neighboring complex molecules.[124]

The X-ray crystal structure of [Mn(chedam)(bpy)]·H$_2$O is shown in Figure 22. The structure contains monomeric [Mn(dicarboxylate)(bpy)H$_2$O] units in which the Mn(II) ions have very irregular six-coordinate geometry. This arrangement is imposed by the small bite angles of the diacid (O11- Mn-O15, 143.44(4)°) and bpy (72.19(5)8) ligands. Individual molecules are linked into chains running parallel to the c axis by hydrogen bonding involving the coordinated water molecule, the phenol group and the carboxylate groups and the chains are linked by π-π stacking of the bpy ligands parallel to the *b* axis.[124]

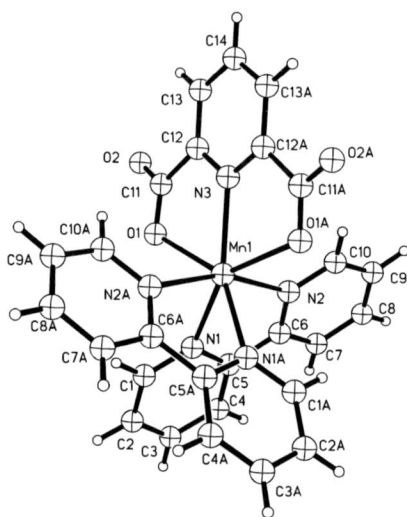

Figure 21. A diagram of [Mn(dipic)(bpy)]. (Reproduced by permission from reference 124).

Iron, in various oxidations, is a very popular element which has been reported to be coordinated by dipicolinic acid.[19, 23-25, 125-132] One unique structure, [(dipic)$_2$(Hdipic)$_2$Fe$^{II}_3$(OH$_2$)$_4$], is reported in the literature.[23] In this unique structure, [(dipic)$_2$(Hdipic)$_2$Fe$^{II}_3$(OH$_2$)$_4$] crystallizes in the centrosymmetric space group *P2$_1$/n.* The asymmetric unit contains half a molecule located on a crystallographic inversion center on Fe(1) as shown on Figure 23 and a crystallization water molecule in general position. The structure consists of two [(dipic)(Hdipic)Fe^{II}] subunits sharing one oxygen atom each with a [O$_2$Fe^{II}(OH$_2$)$_4$] fragment.[23]

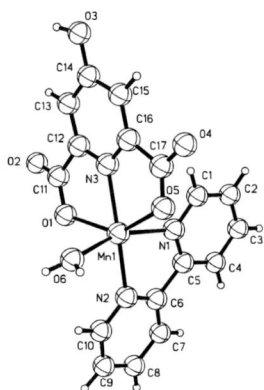

Figure 22. A diagram of [Mn(dipic-OH)(bpy)]. (Reproduced by permission from reference 124).

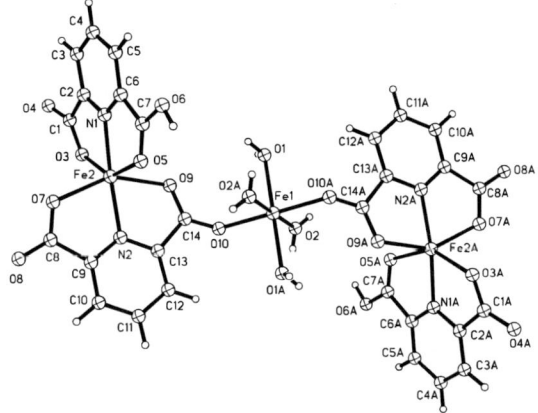

Figure 23. A diagram of [(dipic)$_2$(Hdipic)$_2$Fe$^{II}_3$(OH$_2$)$_4$]. (Reproduced by permission from reference 23).

The coordination sphere around Fe(1) is distorted with an elongation in the Fe(1)-O(110) direction as shown by the distances Fe(1)-O(110) = 2.161(1) Å, Fe(1)-O(120) = 2.085(2) Å, and Fe(1)-O(172) = 2.079(2) Å. It was reported that as observed for [(dipic)Fe^{II}(OH$_2$)$_3$], the shortest bond corresponds to a geometry in which the metal, the oxygen, and the two H atoms are coplanar (sums of the angles around the oxygen: 359° for O(120) and 335° for O(110)).[23] Here, O(110) is not involved in hydrogen bonding, whereas O(120) is linked to the oxygen atom O(111) of an another molecule with O(II)-O(120) = 2.71(3) Å, and O(l1)-H(127) = 1.98(4) Å. The geometry of the two [(dipic)(dipicH)Fe^{II}] subunits is very close to that of [(dipic)$_2$Fe^{II}]$^{2-}$. The acidic H atom is linked by hydrogen bonding to the water molecule with distances H(72)-O(100) = 1.64(4) Å, and O(72)-O(100) = 2.556(3) Å. The protonation of one of the carboxylate groups modifies (see structure II below) the bonding in the complex by the following: (i) a shortening of **a1**, here C(7)-O(71) (1.221(3) Å, to compare with 1.260(3), 1.279(3), and 1.275(3) Å for the similar C-O bonds); (ii) a loss of **b1** (here C(7)-O(72)) double bond character (1.302(3) Å to compare with C(1)-O(12) = 1.232(3) Å, C(11)-O(112) = 1.227(3) Å, and C(17)-O(172) = 1.248(3) Å); (iii) a lengthening of the **c1** bond, Fe(2)-O(71), 2.373(2) Å, whereas the other Fe-O bond (**d1**) ranges from 2.129(3) to 2.148(2) Å. It allows a relaxation of the compression of the Fe-N bond (2.102(2) Å), compared with Fe(2)-N(11) = 2.078(2) Å and less than 2.069(6) Å in [(dipic)$_2$Fe^{II}]$^{2-}$, and the loss of linearity of N-Fe-N' with an angle N(l)-Fe(2)-N(11) = 164.37(8)°.[23]

Reaction of H$_2$dipic and Mohr's salt at pH 6.25 gave the dinuclear species [(dipic)$_2$Fe^{II}(OH$_2$)$_5$], in which a [(dipic)$_2$Fe^{II}] moiety shares one oxygen atom with a [Fe^{II}(OH$_2$)$_5$O] fragment. [23] When the reaction was conducted at pH 5.85, the reaction gave crystals of (NH$_4$)$_2$[(dipic)$_4$(H$_2$dipic)$_2$Fe^{II}$_3$(OH$_2$)$_6$].4H$_2$O, which contains, in the unit cell, both the reactants [Fe^{II}(OH$_2$)$_6$]$^{2+}$ and H$_2$dipic and the product [(dipic)$_2$Fe^{II}][2-23] At pH 5.65, the reaction yielded crystals of [(dipic)$_{10}$(dipicH)$_6$Fe$^{II}_{13}$(OH$_2$)$_{24}$].13H$_2$O, which contains, in the same unit cell, one trinuclear anionic complex ([(dipic)$_2$Fe^{II}]-[Fe^{II}(OH$_2$)$_4$]-[Fe^{II}(dipic)$_2$])$^{2-}$, two doubly protonated cationic dinuclear complexes ((Hdipic)Fe^{II}(Hdipic)]-[Fe^{II}(OH$_2$)$_5$])$^{2+}$, two singly protonated cationic dinuclear complexes [(Hdipic)Fe^{II}(dipic)]-[Fe^{II}(OH$_2$)$_5$]$^+$, and two mononuclear anionic complexes [(dipic)$_2$Fe^{II}][2-23]

For many years now, cobalt complexes with dipicolinic acid have been of interest in coordination chemistry.[38, 56, 133-140] Recently, Moody *et al.*[141] reported the crystal structure of the very first cobalt(II) complex, [Co$_3$Na$_2$(C$_7$H$_2$ClNO$_4$)$_4$(H$_2$O)$_{12}$][Co(C$_7$H$_2$ClNO$_4$)(H$_2$O)$_3$]$_2$.6H$_2$O, with a substituted dipicolinate [4-chloropyridine-2,6-dicarboxylate = dipic-Cl] as ligand. [Co$_3$Na$_2$(C$_7$H$_2$ClNO$_4$)$_4$(H$_2$O)$_{12}$][Co(C$_7$H$_2$ClNO$_4$)(H$_2$O)$_3$]$_2$.6H$_2$O (Figure 24), consists of a centrosymmetric dimer of [Co^{II}(dipic-Cl)$_2$]$^{2-}$ complex dianions [4-chloropyridine-2,6-dicarboxylate = dipic-Cl] bridged by an [Na$_2$Co^{II}(H$_2$O)$_{12}$]$^{4+}$ tetracationic cluster, two independent [Co(dipic-Cl)(H$_2$O)$_3$] complexes, and six water molecules of crystallization.[141] The metals are all six-coordinate with distorted octahedral geometries. The [Co^{II}(dipic-Cl)(H$_2$O)$_3$] complexes are neutral, with one tridentate ligand and three water molecules. The [Co^{II}(dipic-Cl)$_2$]$^{2-}$ complexes each have two tridentate ligands. The [Na$_2$Co^{II}(H$_2$O)$_{12}$]$^{4+}$ cluster has a central Co^{II} ion which is coordinated to six water molecules and lies on a crystallographic inversion center. Four of the water molecules bridge to two sodium ions, each of which have three other water molecules coordinated along with an O atom from the [Co^{II}(dipic-Cl)$_2$]$^{2-}$ complex. In the crystal structure, the various units are linked by O-H•••O hydrogen bonds, forming a three-dimensional network. Two water molecules are disordered equally over two positions.[141]

The synthesis and characterization of Co(II) and Co(III) dipicolinate (dipic^{2-}) complexes were reported by Yang *et al.*[56] X-ray crystallography was carried out on [Co^{II}(H$_2$dipic)(dipic)].3H$_2$O and [Co^{II}(dipic)(μ-dipic)Co^{II}(H$_2$O)$_5$].2H$_2$O. The molecular structure and atom labeling system for [Co^{II}(H$_2$dipic)(dipic)].3H$_2$O is shown in Figure 25. Two nitrogen and four oxygen atoms are coordinated to the cobalt atom resulting in a distorted octahedron. Each of the two tridentate dipic^{2-} ligands coordinates through two oxygen atoms and one nitrogen atom. One of the dipic groups is coordinated as dipic^{2-} and the other as H$_2$dipic, resulting in this complex containing four of the different coordination modes (**a-d**) illustrated in Figure 3 above. The two Co-N bond distances are similar (2.017(3)-2.021(3) Å). Co-O bond distances range from 2.108(2) to 2.222(3) Å (average 2.166 Å).[56] The longest bond is observed in a type c coordination mode (Co-O(2)), and the shortest bond is observed in a type **a** coordination mode (Co-O(12)). The carboxylate group C(6)O(1)O(3) is coordinated to cobalt by the coordination mode shown in **d**, and the carboxylate C(7)O(2)O(4) is coordinated to the cobalt(II) by coordination mode **c**. The carboxylate groups C(16)-O(11)O(13) and C(17)O(12)O(14) are coordinated to the cobalt in coordination modes **b** and **a**, respectively. The identification of both a fully protonated neutral H_2dipic and a dianionic dipic^{2-} ligand in [Co^{II}(H_2dipic)(dipic)] distinguishes this structure from that of the nickel complex, [Ni(Hdipic)$_2$].3H_2O, which contains two monoprotonated Hdipic$^-$ ligands.[49, 50, 53] The structure of [Co^{II}(H_2dipic)(dipic)] exhibits some similarities with the structures of a silver(II) complex ([Ag(Hdipic)$_2$].H_2O)[22] and a copper(II) complex ([Cu(Hdipic)$_2$].3H_2O).[55]

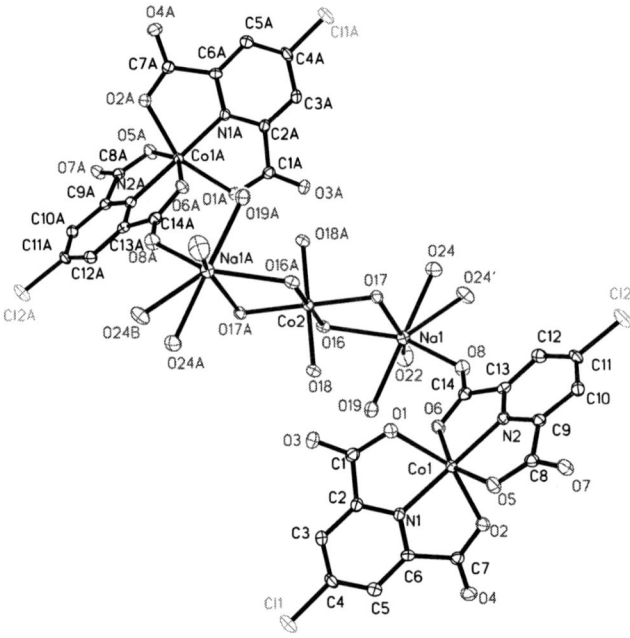

Figure 24. A diagram of [Co$_3$Na$_2$(C_7H_2ClNO$_4$)$_4$(H_2O)$_{12}$] [Co(C_7H_2ClNO$_4$)(H_2O)$_3$]$_2$. 6H_2O. (Reproduced by permission from reference 41).

The formula unit contains three water molecules, one of which is disordered in the crystal structure (O(40)). The other two water molecules form hydrogen bonds with the dianionic dipic ligand. Additional hydrogen bonding between water molecules O(20) and O(40) was found. The major component of O40 (i.e., O40A) is well defined, and hydrogen bonding distances are O(30)•••(4A) (1.54(5) Å), O(20)•••H1 (1.51(5) Å), O(13)•••H(20B) (1.48(5) Å), O(14)•••H(30A) (1.98(5) Å), and O(14)•••H(30B) (1.95(5) Å). Hydrogen bonds are formed between a H-donor and a H-acceptor, and those hydrogen bonds in which the proton originates in [Co^{II}(H$_2$dipic)(dipic)] are shorter than those in which [Co^{II}(H$_2$dipic)(dipic)] is the H-acceptor. An exception to this pattern is observed when the H-acceptor is a carboxylate group bound in coordination mode **b**. The two H$_2$O molecules (O(30) and O(20)) link two complexes through hydrogen bonds. The relevant distances are 1.98(5) Å (H(30A)•••O(14)), 1.54(5) Å (O(30)••••••H(4A)), 1.48(5) Å (H(20B)••••••O(13)) and 1.51(5) Å (O(20)••••••H(1)). All hydrogen atoms except those of the disordered water oxygen atom O40 were found in the electron density map at the expected orientation.

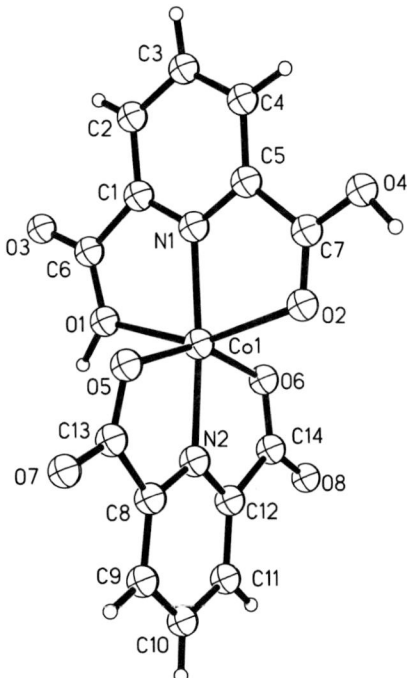

Figure 25. A diagram of [Co^{II}(H$_2$dipic)(dipic)].3H$_2$O. (Reproduced by permission from reference 56).

An ORTEP diagram of [Co^{II}(dipic)(μ-dipic)Co^{II}(H$_2$O)$_5$].2H$_2$O is shown in Figure 26.[56] The two dipic^{2-} ligands are deprotonated in the complex. Both dipic^{2-} ligands are coordinated in a tridentate manner to one cobalt atom; one of the two dipic^{2-} groups also acts as a bridging ligand to the pentaaquo-Co(II) unit. This type of complex has previously been observed in [Zn$_2$(H$_2$O)$_5$(dipic)$_2$].2H$_2$O.[51] Both Co(II) ions exhibit distorted octahedral geometry. The Co-N bond distances are 2.026(2) and 2.033(1) Å, and the Co(1)-O bond distances range from 2.123(1) to 2.225(1) Å (average 2.182 Å). The carboxylate groups are bound in coordination modes **a** and **f**. The Co(2) atom is bound to six oxygen atoms, one from the carboxylate (Co(2)-O(14) 2.097(1) Å) and the other five from water molecules (Co-O from 2.060(2) to 2.180(1) Å, average Co(2)-O 2.097 Å for coordinated water). [56]

Strong hydrogen bonding exists in the crystal structure. Hydrogen bonding between coordinated water (O(23)) and the carboxylate oxygen atoms (O(2) and O(4)) of the dipic^{2-} links the binuclear cobalt molecules to form a one-dimensional chain. The water molecule (O(30)) which hydrogen bonded to three complex molecules stabilizes the chain structure. Relevant distances are 1.81(3) Å (H(25B)••••••O(30)), 1.95(3) Å (H(30A)••••••O(3)), 2.08(3) Å (H(30B)••••••O(4)), 1.92(3) Å (H(23A)••••••O(4)), 2.10(3) Å (H(40B)••••••O(21)). The hydrogen bonds in this structure do not appear to be as strong as those in [Co^{II}(H$_2$dipic)(dipic)].3H$_2$O. [56]

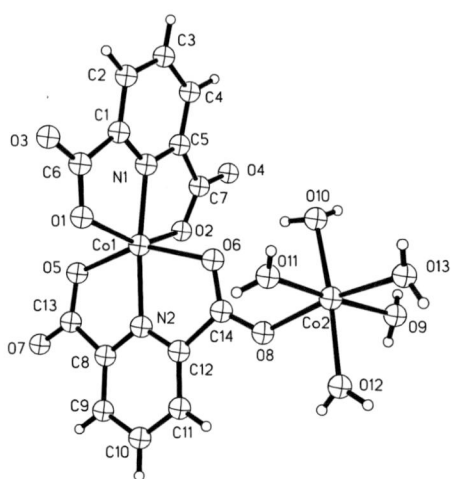

Figure 26. A diagram of [Co^{II}(dipic)(μ-dipic)Co^{II}(H$_2$O)$_5$].2H$_2$O. (Reproduced by permission from reference 56).

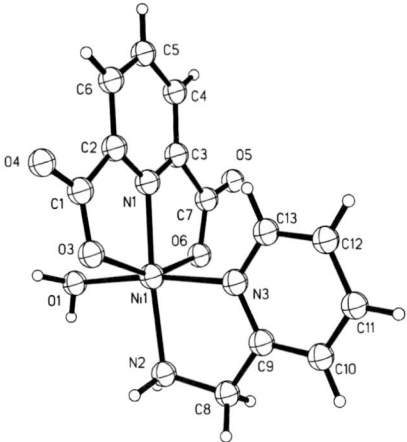

Figure 27. A diagram of [Ni(**1**)(dipic)(OH$_2$)]. (Reproduced by permission from reference 142).

Reaction of nickel(II) or copper(II) acetate with 2-(aminomethyl)pyridine **1** and the dipicolinate anion in aqueous methanol in a 1:1:1 molar ratio resulted in the formation of [Ni(**1**)(dipic)(OH$_2$)] (Figure 27) or [Cu(**1**)$_2$(CH$_3$OH)][Cu(dipic)$_2$] (Figure 28).[142] The nickel complex crystallizes as a highly hydrated structure with intricate hydrogen-bonding interactions, almost protein-like in character; that is elaborated by the complexes being ordered through π-stacking of the 2-(aminomethyl)pyridine ligands and hydrogen bonding in a type of a 'dimer tape' polymer arrangement. The asymmetric unit consists of two independent [Ni(1)(dipic)(OH$_2$)] molecules linked by close hydrogen bonds (Figure xxx), and ten water molecules. There are five fully occupied water molecules (O01, O02, O03, O05, and O06), one fully occupied water, but disordered over five positions (O010), and three half-occupied water molecules (O04, O07, and O08). The nickel centers feature bidentate coordination of **1** and tridentate coordination of dipic^{2-}, and with a coordinated water molecule. The structure exhibit a distorted octahedral *mer*-NiN$_3$O$_3$ geometry, with each axis involving a unique pair of donors (N$_{py}$ and N$_{amine}$; N$_{py}$ and O$_{water}$, O$_{COO}$ and O$_{COO}$).[142]

The two independent nickel complexes in the structure exhibit very similar distances and angles around the nickel metal centre. The two independent complexes are linked by hydrogen bonding interactions between the coordinated water of one nickel and a carboxylate oxygen of the other [O1•••O201 2.718(3) Å], reciprocated by the water and a carboxylate oxygen of the other center [O2•••O101 2.708(3) Å] were also reported.[142] Each coordinated water is also hydrogen bonded to the carboxylate oxygen of a symmetry related molecule [O1-O112 2.682(3) Å for Ni1 and O2•••O212 2.682(3) Å for Ni2]. Distortion in the octahedron around the nickel is also clear in figure 27. The planar tridentate chelation of dipic^{2-} leads to the O212-Ni2-O201 angle being reduced to 155.13(7)°, with intrachelate angles such as N205-Ni2-O212 at 77.72(8)° severely compressed compared with the relatively unstrained angle of the coordinated water N205-Ni2-O2 of 92.20(9)°. The other meridional angle involving **1** and the coordinated water, N220-Ni2-O2 of 172.42(9)°, also reflects some folding back, but not to the same extent as observed with coordination of dipic^{2-}. The Ni-N(pyridine) distance in the tridentate dipic^{2-} is significantly compressed [average 1.981(2) Å] compared with the distance of chelate **1** [average 2.073(2) Å]. The distances of Ni-O vary from an average of 2.134(2) Å for the carboxylate oxygens in dipic^{2-} to 2.085(2) Å for H_2O, despite the ionic nature of the former oxygen expecting a smaller distance. Both effects reflect the small bite angle of dipic^{2-} for the relatively large Ni(II) cation. The previously reported [Ni(Hdipic)$_2$].3H$_2$O complex exhibits an almost regular octahedral geometry,[49, 53] with the Ni-O distances in the range 2.10-2.21 Å, the longer distances being associated with the protonated acid group; the Ni-O (carbaoxylate) distances are comparable with the current Ni(II) complex, as are the Ni-N distances.[142]

The distortions due to the bidentate ligand **1** are reduced due to longer intra-ligand distances in the arm pendant to the pyridine ring compared to the case for dipic^{2-} and a longer preferred Ni-N(amine) distance compared to Ni-O(carboxylate); nevertheless, the N220-Ni2-N213 angle, for example, is compressed somewhat, at 80.96(9)°. Throughout the structure, planarity of the pyridine rings is maintained, albeit with some minor distortions reflected in ellipticity of probability ellipsoid for some ring atoms distant from the coordination sites.[142]

The copper(II) complex asymmetric unit contains two types of Cu(II) ions and three solvent methanols, consisting of independent cations [Cu(1)$_2$(CH$_3$OH)]$^{2+}$ and independent anions [Cu(dipic)$_2$]$^{2-}$ (Figure 28). The oxygen of the two methanols (O1 and O01) are coordinated strongly and weakly to Cu2. The other methanol (O02) is involved in receiving a hydrogen bond from the amine (N408) of one of the Cu1 ions and donating a hydrogen bond to the carboxylates (O101 and O201) of a Cu(II) ion. There are many other close contacts throughout the system, the hydrogen bonding along the direction of Jahn-Teller elongation in the two distinct Cu(II) ions link these and one of the unbound methanols in a polymer-like "ribbon" that apparently contributes significantly to the driving force for formation of the isolated solid.[142]

The crystal structure of [Cu(dipic)$_2$]$^{2-}$ has been described before, but with a different cation.[143] Bond distances and angle in the current complex are very similar to the reported complex. The complex is isolated as a neutral [Cu(dipic)(H$_2$dipic)].xH$_2$O form, where two protonated carboxylate groups of one ligand occupy elongated apical positions of the Jahn-Teller distorted octahedron. Both monohydrate and trihydrate have been reported in the past,[144] with Cu-N distances in the range 1.901-1.907 (dianion) and 1.995-2.003 Å (diacid), Cu-OCO⁻ 2.008-2.063 Å and Cu-OCOH 2.302-2.465 Å. In the present structure, Cu-N distances vary from 1.896 Å, comparable to the dianion of the neutral complex, to a longer 1.9555 Å where the N is a component of the tridentate ligand with the Jahn-Teller elongated Cu-carboxylates, reflecting a compensation necessary to avoid serious angle strain in the carboxylate arms of the planar ligand. This Cu-N distance is still shorter than the reported when the Jahn-Teller elongated distances are also protonated. The Cu-OCO⁻ distances of 2.044 and 2.049 Å are comparable to those in the literature, whereas the Jahn-Teller elongated distances of 2.336 and 2.353 Å lie at the shorter end of the range reported for Jahn-Teller elongated and protonated carboxylates.[142]

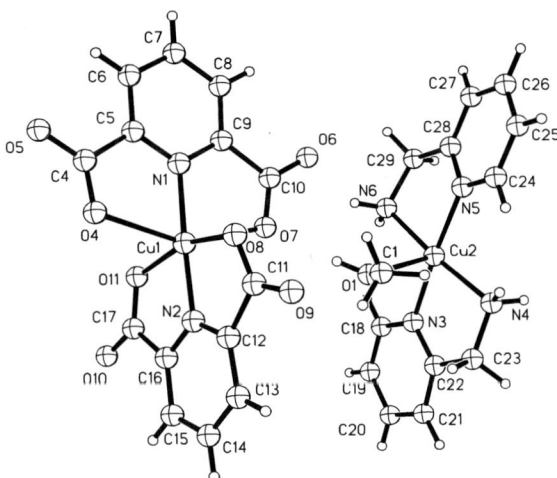

Figure 28. A diagram of [Cu(1)$_2$(CH$_3$OH)][Cu(dipic)$_2$]. (Reproduced by permission from reference 142).

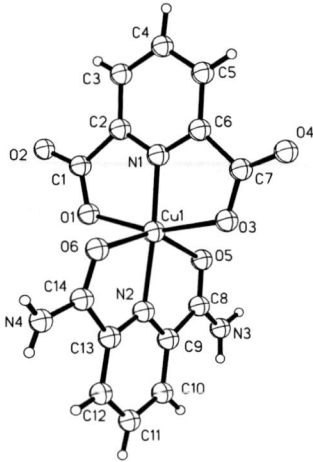

Figure 29. A diagram of [Cu(dipic)(pdcam)].2H$_2$O. (Reproduced by permission from reference 145).

The compound (2,6-pyridine-dicarboxamide) (2,6-pyridine-dicarboxylato) copper(II) dihydrate ([Cu(dipic)(pdcam)].2H$_2$O), was synthesized by the reaction of Cu$_2$(CO$_3$)(OH)$_2$, H$_2$dipic, and pdcam in an aqueous-ethanol medium.[145] [Cu(dipic)(pdcam)].2H$_2$O consists in elongated octahedral molecules where the copper(II) atom exhibits a coordination surrounding N$_2$O$_2$ + O$_2$, type 4 + 2 (Figure 29). The four closest donor atoms are the N-pyridine and two O-carboxylate donors of the tridentate dipic^{2-} ligand and the N-pyridine atom from the pdcam ligand.[145] Both primary amide O atoms from pdcam occupy the trans-apical positions of the Cu(II) coordination polyhedron, with bond lengths ~2.30 Å and define the lowest trans-angle of [Cu(dipic)(pdcam)].2H$_2$O (~150°). The trans-angle O(dipic)–Cu–O(dipic) also has a significantly low value (~160°). Both O–Cu–O trans angles of [Cu(dipic)(pdcam)].2H$_2$O reveal the rather rigid structures of such tridentate ligands, which are roughly planar (within 0.018 (dipic) or 0.079 (pdcam) Å. In contrast, the trans-angle N–Cu–N is very close to ~180° and the dihedral angle defined by the mean planes of dipic^{2-} and pdcam ligands is 88.4°, showing that they fall perpendicular. The lowest planarity of pdcam seems related to the implication of all N(amide)–H bonds in the 3-D H-bonding network of the crystal, with O carboxylate or water acceptor atoms. Stacking π,π-interactions between aromatic rings of the ligands are missing. On the other hand, all Cu–N and Cu–O bond lengths of [Cu(dipic)(pdcam)].2H$_2$O follow the ligand order dipic^{2-} < pdcam, probably because dipic^{2-} is a divalent anion whereas pdcam is a neutral ligand. The mer-NO_2(equatorial) conformation preferred by $dipic^{2-}$ imposes to pdcam a mer-N(equatorial) + O_2(apical) conformation in the elongated octahedral Cu(II) coordination polyhedron, thus featuring its conformational flexibility.[145]

The molecular structures of [Cu(dipic)(4dmapy)] (where 4dmapy = 4-dimethylaminopyridine) and [Cu(dipic)(nmim)(H_2O)$_{0.5}$] (where nmim = N-methylimidazole) are shown in Figure 30.[146] The structure of [Cu(dipic)(4dmapy)] is square planar with the 4dmapy ligand coplanar with the CuN_2O_2 coordination plane, dihedral angle 1.3°. The Cu-O bond lengths of ~2 Å are normal, whereas the Cu-N distances are short. The four coordinate atoms lie slightly below the CoN_2O_2 best plane, N(1) 0.0174, N(2) 0.0184, O(1) 0.0040 and 0(3) 0.0042 Å, whilst the copper atom is above the plane by 0.0440 Å. The dimensions of the dipic ligand with bite angles of ~80° are in normal ranges. 4dmapy is mainly of the canonical form **1b**, as is evident by the short distances of C(10)-N(3), C(8)-C(9) and C(11)-C(12), and the essential coplanarity of the dimethylamine moiety and the pyridine ring. This is attributable to the delocalization of the amine lone pair with the pyridine nucleus.[146]

[Cu(dipic)(nmim)(H_2O)$_{0.5}$] consists of two distinct asymmetric copper units, a Cu(1) square planar unit [Cu(dipic)(nmim)] and a Cu(2) square pyramidal unit [Cu(dipic)(nmim)(H_2O)]. Both units contain a CuN_2O_2 square plane formed by the dipicolinate and the nmim ligand, where the Cu-N bond lengths are short and the Cu-O distances are normal. The $dipic^{2-}$ bite angles are close to 80°. In the planar unit, the Cu(II) ion (0.0336 Å), O(1) (0.0034 Å] and O(3) (0.0031 Å) are slightly above the CuN_2O_2 best plane, while N(1) (-0.0208 Å) and N(2) (-0.0194 Å) lie slightly below. The dihedral angle between the imidazole nucleus and the CuN_2O_2 coordination plane is 6.8°. In the Cu(2) unit, N(4) (0.0688 Å), N(5) (0.0686 Å), O(5) (0.0031 Å), and O(7) (0.0044 Å) are slightly below the CuN_2O_2 best plane, and the copper(II) ion above (0.1449 Å). The dihedral angle between the imidazole nucleus and the CuN_2O_2 coordination plane is 13.9°. Hydrogen bonds were observed between the O(9) and O(4a) atoms (2.295 Å). The dimensions of the dipic^{2-} and nmim ligands are in the normal ranges.[146]

There are two complexation modes for dipicolinate copper(II) complexes; first, the dipic^{2-} binds strongly in the equatorial plane with a short Cu-N bond of ~1.9 Å and Cu-O bond of ~2.0 Å, as in [Cu(dipic)(H$_2$O)$_2$]$_2$,[54] secondly, the dipic^{2-} bonds perpendicularly to the equatorial plane with Cu-N of ~2.0 Å and Cu-O of ~2.4 Å, as in [Cu(dipic)(tpy)].[147] The [Cu(dipic)$_2$] moiety in [Cu$_2$(dipic)$_2$(bpy)$_2$]·4H$_2$O[148] comprises both modes. The dipic ligands in both complexes belong to the first category, i.e., the dipic^{2-} ligands bind strongly in the equatorial coordination plane.[146]

The self-assembly of 4-hydroxypyridine-2,6-dicarboxylic acid (H$_3$CAM) and dipicolinic acid (H$_2$dipic) with Zn(II) salts under hydrothermal conditions gave two novel coordination polymers ([Zn(HCAM)]·H$_2$O)$_n$ and ([Zn(dipic)(H$_2$O)$_{1.5}$])$_n$.[149] Figure 31 shows a portion of an ORTEP diagram for ([Zn(dipic)(H$_2$O)$_{1.5}$])$_n$.[149]

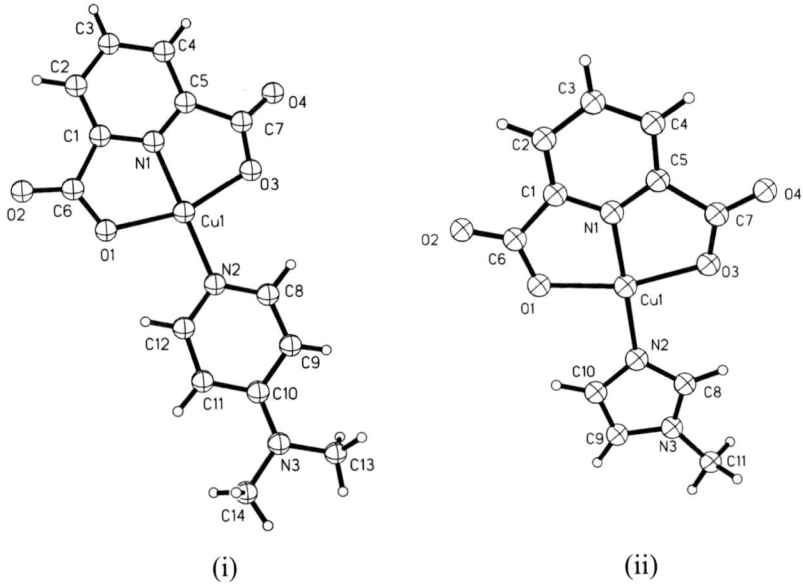

Figure 30. Diagrams of [Cu(dipic)(4dmapy)] (i) and [Cu(dipic)(nmim)(H$_2$O)$_{0.5}$] (ii). (Reproduced by permission from reference 146).

Coordination Chemistry of Dipicolinic Acid and Its Analogues

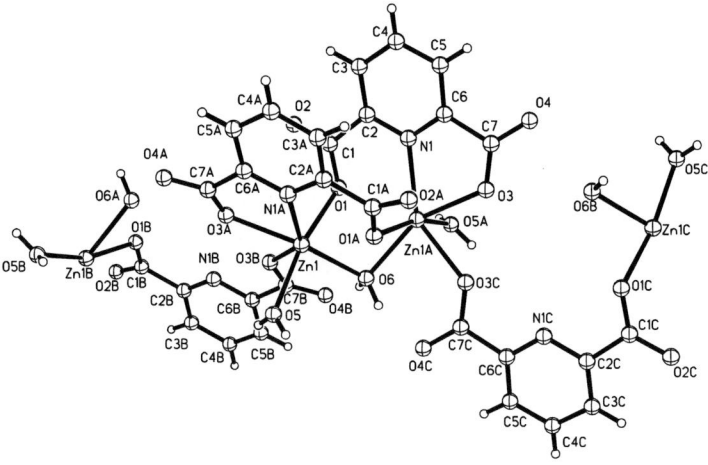

Figure 31. A diagram of ([Zn(dipic)(H$_2$O)$_{1.5}$])$_n$. (Reproduced by permission from reference 149).

Three novel compounds were obtained as isomorphous ionic crystals containing [M(dipicH$_2$)(OH$_2$)$_3$]$^{2+}$ and [Ce(dipic)$_3$]$^{2-}$ ions.[150] The tri-capped trigonal prismatic Ce(IV) chelate has also been observed in Ce(IV)-alkaline earth-dipic complexes [151-154]. However, in all these cases some of the free oxygen atoms of the chelate coordinate to M(II) ions forming chains and networks. In the present cases there is no coordination link between the anions and cations. Instead, the free oxygen atoms form H-bonds with carboxylic groups and coordinated water molecules of the cation.[150]

All the three cations, [M(dipicH$_2$)(OH$_2$)$_3$]$^{2+}$, have very similar 4+2 coordination polyhedra. Figure 32 shows an ORTEP diagram for [Zn(dipicH$_2$)(OH$_2$)$_3$][Ce(dipic)$_3$]. The equatorial plane contains the nitrogen atom of dipicH$_2$ and the O atoms of the three water molecules, while the axial positions are occupied by the carbonyl oxygen atoms of dipicH$_2$. The major difference is that the axial elongation is significantly more in the case of the copper complex. The average equatorial and axial distances (Å) are 2.026(3), 2.175(3) (Ni); 1.984(2), 2.344(2) (Cu); 2.057(1), 2.236(1) (Zn). It would appear that the expected vibronic effects are superimposed on the steric requirements of the ligand to produce the observed static structure in the case of the copper complex. The average off-axis deviation (°) of the axial ligand atoms are 13.5 (Ni) and 15.1 (Cu, Zn). The equatorial atoms including the metal atoms are very nearly coplanar in all cases with a largest deviation of the atoms (Å) from the mean plane 0.038(3) (Ni), 0.062(2) (Cu) and 0.111(2) (Zn). The deviations amount to a very slight tetrahedral distortion with the following inter planar angles (°): 3.3(2) (Ni); 4.6(1) (Cu); 8.2(1) (Zn).[150]

A very interesting report was published on the self-assembly of polyoxometallate clusters into a 3-D heterometallic framework via covalent bonding: synthesis, structure and characterization of $Na_4[Nd_8(dipic)_{12}(H_2O)_9][Mo_8O_{26}].8H_2O$,[155] but prior to that the crystal structures of [$MoO_2(dipic)(L)$] (L = DMF, DMSO, OPPh$_3$) were reported.[156] Single-crystal X-ray diffraction studies on [$MoO_2(dipic)(L)$] (L = DMF, DMSO, OPPh$_3$) revealed that the three complexes have a closely related structure, similar to that of [$MoO_2(dipic)(DMF)$] (see Figure 33).[156] The molybdenum atom adopts the expected distorted octahedral geometry, which is determined by the usual way of coordination for the dipicolinate ligand (planar tridentate) to the angular *cis*-MoO$_2$ moiety. Selected bond lengths and angles are listed in Table 8.[156] The bond lengths for Mo-O(5) and Mo-O(6) are not identical, the former being shorter. While the Mo-O(6) distance, involving the oxygen *trans* to N(1), is nearly identical for all the complexes, the Mo-O(5) distance increases with the basicity of the *trans* ligand.[156]

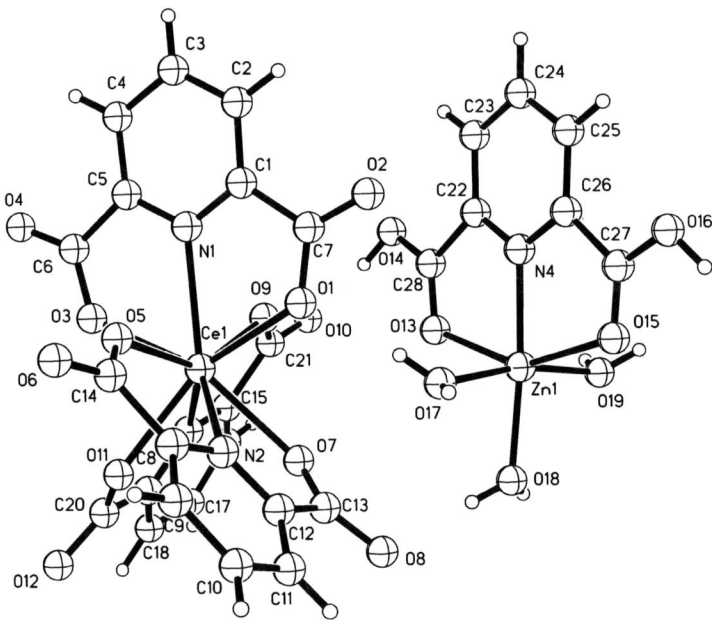

Figure 32. A diagram of [$Zn(dipicH_2)(OH_2)_3$][$Ce(dipic)_3$]. (Reproduced by permission from reference 150).

Table 8. Selected bond lengths (Å) and angles (°) for [MoO$_2$(dipic)(DMF)], [MoO$_2$(dipic)(DMSO)], and [MoO$_2$(dipic)(OPPh$_3$)]

	[MoO$_2$(dipic)(DMF)]	[MoO$_2$(dipic)(DMSO)]	[MoO$_2$(dipic)(OPPh$_3$)]
Mo(1)-O(5)	1.674(2)	1.696(2)	1.687(2)
Mo(1)-O(6)	1.702(2)	1.703(2)	1.700(2)
Mo(1)-O(1)	2.009(2)	2.008(2)	2.003(2)
Mo(1)-O(4)	2.010(2)	2.017(2)	2.000(2)
Mo(1)-N(1)	2.190(2)	2.198(2)	2.205(2)
Mo(1)-O(7)	2.308(2)	2.312(2)	2.247(2)
O(4)-C(7)	1.320(4)	1.335(3)	1.326(3)
O(1)-C(1)	1.325(4)	1.334(3)	1.325(3)
O(7)-X	1.233(4)	1.544(2)	1.502(2)
O(5)-Mo(1)-O(6)	105.5(1)	104.7(1)	104.6(1)
O(1)-Mo(1)-O(4)	144.68(9)	144.81(7)	145.43(8)
O(6)-Mo(1)-O(4)	106.5(1)	101.82(8)	103.86(9)
O(6)-Mo(1)-O(1)	101.8(1)	106.90(9)	104.73(8)
N(1)-Mo(1)-O(4)	72.48(8)	72.56(7)	72.95(8)
N(1)-Mo(1)-O(1)	72.79(8)	73.12(7)	73.03(7)
O(5)-Mo(1)-O(4)	95.8(1)	94.49(9)	72.95(8)
O(5)-Mo(1)-O(1)	96.6(1)	97.40(8)	95.79(9)
O(6)-Mo(1)-O(7)	82.3(1)	83.86(8)	85.04(8)
N(1)-Mo(1)-O(7)	71.76(8)	73.97(6)	71.31(7)
N(1)-Mo(1)-O(5)	100.5(1)	97.40(8)	99.06(9)
O(5)-Mo(1)-O(7)	172.0(1)	171.3798)	170.36(9)
N(1)-Mo(1)-O(6)	153.9(1)	157.58(8)	156.35(9)

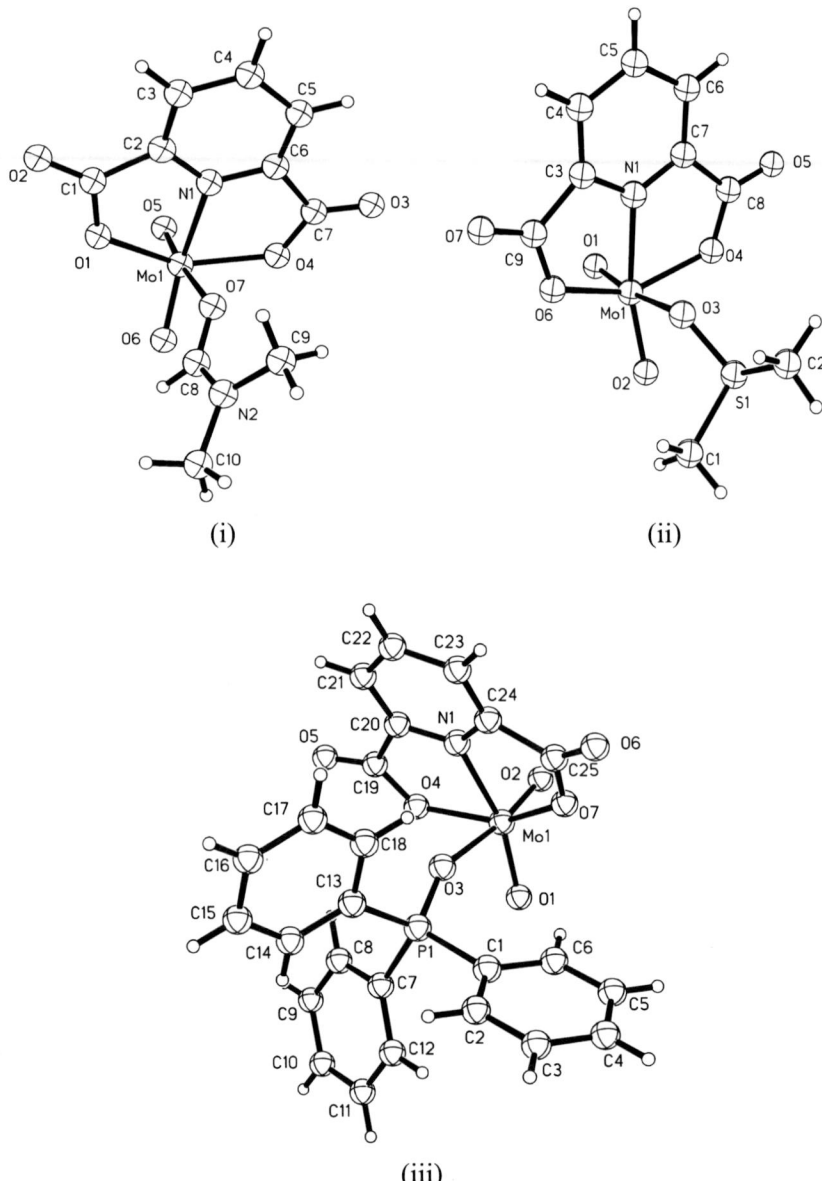

Figure 33. Diagrams of [MoO$_2$(dipic)(DMF)] (i), [MoO$_2$(dipic)(DMSO)] (ii), and [MoO$_2$(dipic)(OPPh$_3$)] (iii). (Reproduced by permission from reference 156).

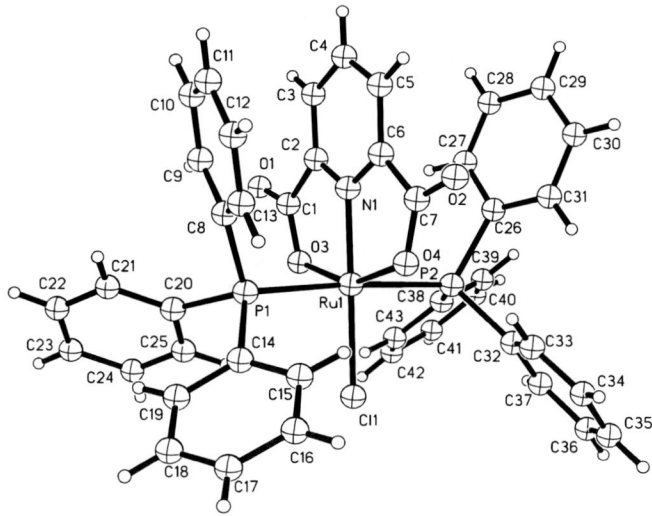

Figure 34. A diagram of [Ru(dipic)(PPh$_3$)$_2$Cl]. (Reproduced by permission from reference 157).

New hexa-coordinated Ru(III) complexes of the type [Ru(dipic)(EPh$_3$)$_2$X] (where X = Cl, Br; E = P, As) were synthesized by reacting dipicolinic acid with the appropriate starting complexes [RuX$_3$(EPh$_3$)$_3$].[157] The ligand behaves as tridentate dibasic chelate. A diagram of [Ru(dipic)(PPh$_3$)$_2$Cl] is shown in Figure 34. The N(1)–Ru(1)–Cl(1) bond angle is 178.95(9)° showing that Cl atom lies trans to ring nitrogen. The bite angles around Ru(III) are N(1)–Ru(1)–O(4) = 77.42(10)°; N(1)–Ru(1)–O(3) = 77.28(9)°; O(4)–Ru(1)–Cl(1) = 103.60(7)°, and O(3)–Ru(1)–Cl(1) = 101.70(6)°, summing up the in-plane angle to be exactly 360°. This shows the high planarity of the Cl and O, N, O donor atoms of dipicolinic acid. This was further supported by the other cis angles as reported.[157] The acid occupies the equatorial plane around the Ru(III) octahedron, along with Cl. The bond angle P(1)–Ru(1)–P(2) = 175.62(3)° shows that the two PPh$_3$ groups are trans to each other occupying the axial positions. The two Ru–P bonds are slightly bent away from the dipicolinic acid towards the Cl atom which is evident from the P(1)–Ru(1)–Cl(1) bond angle is 88.84° which is smaller than P(1)–Ru(1)–N(1) = 91.41° and P(2)–Ru(1)–N(1) = 92.51°.[157] The Ru–P, Ru–O, Ru–N and Ru–Cl bond lengths found in the complex agree well with that reported for similar ruthenium complexes.[157]

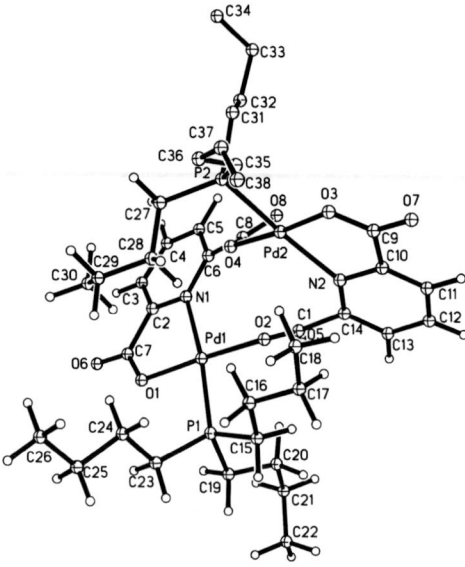

Figure 35. A diagram of [Pd(dipic)(PBu$_3$)]$_2$. (Reproduced by permission from reference 28).

The reactions of [Pd(acac)$_2$] or [Pd(OAc)$_2$]$_3$ with dipicolinic acid in acetonitrile produced [Pd(dipic)(NCMe)], which was used to synthesize [Pd(dipic)(PBu$_3$)]$_2$.[28] The X-ray structure and numbering scheme for [Pd(dipic)(PBu$_3$)]$_2$ are shown in Figure 35. The X-ray determination confirms the structure proposed on the basis of spectroscopic data.[28] The molecule is a dimer in which each dipic^{2-} ligand is coordinated to one palladium in a chelate manner through one carboxylate oxygen atom and the pyridine nitrogen, while the other carboxylate group, twisted out of the pyridine plane, is bonded to the second palladium atom of the dimer. The two bridges are arranged in a complementary head-to-tail manner. The resulting coordination around each palladium is distorted square planar with the remaining fourth coordination site filled by the phosphorus atom of the tributylphosphine.[28]

In related complexes the pyridine ring and the chelating carboxylate are nearly coplanar with the coordination plane.[26, 27, 158] However, in [Pd(dipic)PBu$_3$]$_2$ the pyridine ring is twisted by 20.0(2)° (for that on Pd(1)), or 12.2(2)° (for that on Pd(2)), with respect to the coordination planes. The deviation from coplanarity of the chelated carboxylate groups with respect to the pyridine ring is reflected in the torsion angles O(1)-(11)-C(12)-N(1) = -3(1)°, and O(2)-C(21)-C(22)-N(2) = -7(1)°. The bridging carboxylate groups are rotated with respect to the pyridine ring by 36.4(4) and 42.1(4)°. The C-O distances and angles for the chelate and bridged carboxylate groups are very similar since both carboxylates are coordinated. The dipic^{2-} ligands are oriented in such a way that their aromatic rings are very nearly perpendicular to one another (dihedral angle 84.4(2)°), whereas the dihedral angle between the coordination planes of the two palladium atoms is 69.76(8)°. The phosphine seems to exert a marked influence on the *trans* Pd-N distances (2.140(5) and 2.139(6) Å). They are noticeably longer than the distances found in [Pd(dipic)Br]$^-$ (2.034(8) Å)[159] or in dipicolinate complexes of platinum (1.88-2.03 Å)[26] and are similar to those found in the series of dimers [Pd(μ-η^2-pySN,S)Cl(PR$_3$)]$_2$ (PR$_3$ = PMe$_3$, PMe$_2$Ph, and PMePh$_2$),[160] all of them having the PR$_3$ ligand in position *trans* to the N atom, which are in the range of 2.124(6)-2.137(9) Å.

The tendency of the dipicolinate complexes to dimerization is absent in the related complex [Pd(pdtc)PPh$_3$], as determined by ^{1}H NMR spectroscopy.[159] A plausible explanation for this different behavior is suggested by the structural features of both families. For [Pd(pdtc)Br]$^-$ the chelating cycle is relatively free of strain (N-Pd-S) 86.44(6)°), whereas for [Pd(dipic)PBu$_3$]$_2$ the chelating carboxylate groups have a smaller bite angle (N(1)-Pd(1)-O(1) = 81.0(2)°; N(2)-Pd(2)-O(2) = 81.2(3)°). This is related to the shorter C-O distance (1.299(9) and 1.27(1) Å) compared to the C-S distance (1.711(9) Å, as well as by the longer Pd-N distance in [Pd(dipic)PBu$_3$]$_2$. Consequently the chelating cycles in the dipicolinate derivatives are more strained and have a higher tendency to relieve this strain by forming dimers (with only one strained cycle per palladium) rather than monomers (with two strained cycles per palladium), specially when the ancillary ligand induces a long Pd-N bond.[28]

A mixed Co-Ag complex with H$_2$dicpic, AgCo(dipic)$_2$, was synthesized under hydrothermal conditions.[161] X-ray structural analysis reveals that the asymmetric unit consists of one Ag(I) ion, one Co(III) ion and two dipic^{2-} ligands. The complex crystallizes in the high symmetry space group $I4_1/a$. The Co(III) metal center has a distorted octahedral environment, with four *anti*-O(2) atoms of two dipic^{2-} ligands in the equatorial plane and two N atoms in axial positions (figure 36). The two $^{2-}$ ligands are perpendicular to each other. The four *syn*-O(1) atoms of dipic^{2-} are bonded at a distance of 2.4679(16) Å to Ag(I) to form an AgO$_4$ tetrahedron. All Co(III) and Ag(I) ions lie in a plane with a shortest Co···Ag contact of 4.795 Å. A search of the Cambridge Structural Database (version 5.26 [162]) revealed that the [M(dipic)$_2$]$^{n-}$

building block has been reported many times, but in those structures units are extended by interactions such as hydrogen bonding and π-stacking [57, 163, 164]. The present structure is the first example of 2-D sheet structure with 4^4 topology. Each carboxylate group associated with two [Co(dipic)$_2$]$^-$ ions links two Ag(I) ions to give a 4^4 tessellated 2-D net of Co$_2$Ag$_2$ groups with a Co···Ag separation of 4.975 Å.[161] Significant π-stacking involving pyridine rings is evident between the layers. The stacking propagates along the *a* and *b* axes in a decussated fashion (figure 4). This places pyridine rings of neighboring layers on top of one another with a separation of 3.10 Å (centroid–centroid = 3.50 Å, slippage angle = 27.5°). The rest of the layer contents stack by filling alternate bumps and hollows in adjacent layers.[161]

Although the π-stacking exhibits an offset face-to-face motif, the short separation distance indicates a strong π-π interaction.[161]

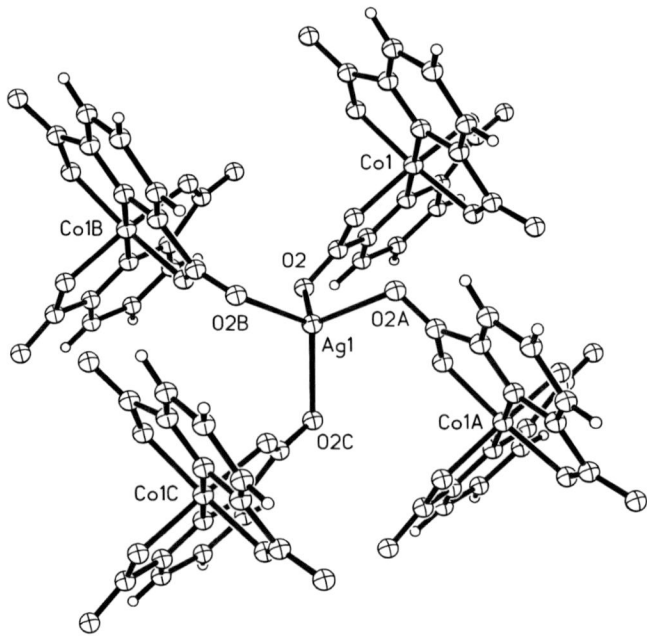

Figure 36. A diagram of the mixed-metal Co(III)-Ag(I) complex. (Reproduced by permission from reference 161).

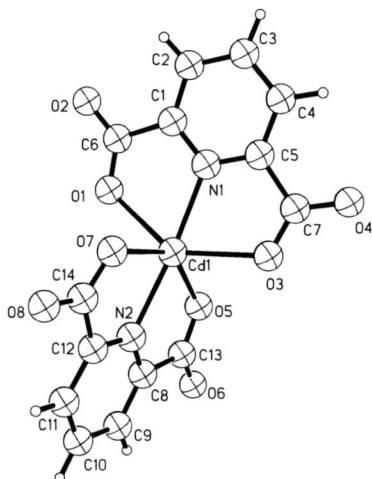

Figure 37. A diagram of the [Cd(dipic)$_2$]$^{2-}$ anion. (Reproduced by permission from reference 165).

A novel 1:2 proton transfer self-associated compound **LH2**, (GH$^+$)$_2$ (dipic^{2-}), was synthesized from the reaction of dipicolinic acid and guanidine hydrochloride, (GH$^+$)(Cl$^-$).[165] The reaction of **LH$_2$** with cadmium(II) iodide in 2:1 molar ratio in water produced (GH)$_2$[Cd(dipic)$_2$]·2H$_2$O The molecular structure of the complex with atom numbering scheme and the crystal packing diagram are shown in Figure 37.[165] Important features of the crystal structure are the presence of an anionic complex [Cd(dipic)$_2$]$^{2-}$, cationic counter ion GH$^+$ and complexation of two pyridine dicarboxylates as tridentate ligands. The metal ion is hexacoordinated by two nitrogen atoms N(1), and N(9) and four oxygen atoms O(1), O(3), O(5) and O(7) of carboxylato groups of two dipic^{2-}. Fig. 2 shows that the Cd^{II} metal centre is located in the center of a distorted octahedral arrangement. The N(1)-Cd-N(9) angle shows deviation from linearity, 167.64(6)°. The C(6)-N(1)-Cd-O(7), C(6)-N(1)-Cd-O(5), C(10)-N(9)-Cd-O(3) and C(10)-N(9)-Cd-O(1) torsion angles are 101.77(14)°, -77.47(15)°, -81.78(14)°, and 102.33(14)°, respectively, indicating that two dianionic dipic^{2-} units are almost perpendicular to each other. Another characteristic solid state structural feature of this complex is dictated by the presence of guanidine fragment as a strong base that fully deprotonates pyridinedicarboxylic acid. This can lead to the formation of a Cd^{II} complex in which ion-pairing, metal-ligand coordination and intensive hydrogen-bonding play important roles in the construction of its three dimensional supramolecular network. It has been shown that 2,6-pyridinediamine (pyda) has key a role in the construction of a number of metal complexes of this proton transfer compound.[166-168] In the case of the Cd^{II} complex, for instance, a centrosymmetric dinuclear Cd^{II} complex containing pydaH$^+$ cation is prepared in which the two metal fragments are linked *via* the central four-membered Cd_2O_2 ring and each cadmium atom is in the center of a distorted pentagonal bipyramid.[168] As it is clear, the substitution of the pydaH$^+$ cation with GH$^+$ significantly changes the structure of the resulting Cd^{II} complex. These data confirm the influence of the counter ion in the complexation as well as the ligation and stereochemistry of the complex.[165]

An important feature in both **LH$_2$** and (GH)$_2$[Cd(dipic)$_2$]·2H$_2$O complex is the presence of GH$^+$ cation with D_{3h} symmetry. The three C-N bond lengths in GH$^+$ are very close in values being 1.329(3) Å, 1.328(3) Å and 1.328(3) Å. These bond distances that are intermediate between C-N and C=N bonds indicate the equally distributed positive charge on three guanidinium nitrogen as expected. The angles N-C-N are all close to 120°. Each GH$^+$ cation adopts a planar structure and the multiple-electron delocalized π bond is thus formed. It is interesting to note that the C-N bond lengths of various types of guanidinium containing complexes reported to date vary in a range from 1.279 to 1.402 Å.[165] Extensive hydrogen bondings between carboxylate, GH$^+$, and water molecules throughout the lattice of Cd^{II} complex play important roles in stabilizing the crystal. The range of D-H•••A angles and the H•••1468; A and D•••1468; A distances indicate the presence of strong hydrogen bonding in the Cd^{II} complex lattice.[165]

A unique tin-containing complex was recently solved by X-ray crystallography.[169] X-ray diffraction investigation of [(C$_2$H$_5$)$_3$NH][Me$_3$Sn(dipic)] (Figure 38) shows that it is not a infinite polymeric chain but an oligomer with three trimethyltin centers linked by two dipic^{2-} ligands and bounded by two water molecules coordinating the terminal tin atoms.[169] In the case of [(C$_2$H$_5$)$_3$NH][Me$_3$Sn(dipic)], both the trinuclear oligomeric anion and the counter cation are subject to crystallographically imposed twofold axial symmetry. Thus Sn(2) and C(11) of the anion and N(2) and C(16) of the cation are in special positions on a twofold axis, while all other atoms of the asymmetric unit, including the methylene group involving C(15) which is disordered over two symmetry related sites of equal occupancy, are in general positions. In [(C$_2$H$_5$)$_3$NH][Me$_3$Sn(dipic)], each dianion bridges two Sn centers via only one O atom derived from the monodentate carboxylate moiety. Owing to the coordinated water molecules, the environments of the two terminal Sn atoms are different to another Sn atom. As a result of the bidentate mode of coordination of the dicarboxylic acid, each Sn center is five-coordinate and exists in trigonal bipyramidal geometry with the O (derived of water and carboxyl groups, respectively) atoms occupying the axial sites [Sn(2)–O(3) 2.266(4), Sn(2)–O(3^1^) (symmetry code: -x + 3/2, -y + 5/2, z) 2.266(4), Sn(1)–O(1) 2.179(3), Sn(1)–O(5) 2.396(4) Å and O(3)–Sn(2)–O(3^1^) 165.9(2)°, O(1)–Sn(1)–O(5) 177.38(14)°].[169]

The C–O bond distances [C(1)–O(2) 1.225(6) Å, C(7)–O(4) 1.217(7) Å] associated with the non-coordinating O atoms are significantly shorter than the coordinating C–O bond distances [C(1)–O(1) 1.278(6) Å, C(7)–O(3) 1.252(6) Å]. The intramolecular Sn(2)•••O(4) of 3.542 Å, is not indicative of bonding interactions between these atoms. Although not involved in coordination to tin, the O(4) atom form significant intermolecular contacts in the crystal lattice.[169]

Through the coordinated water molecule to the terminal tin atoms of the polymer, the polymers are associated with each other via hydrogen bonds to the pendant O atoms of carboxyl groups and the N atoms derived of the pyridine ring, so that a 2-D network is formed. The distances of hydrogen-bonding, O(5)–H•••O(2^1^), O(5)–H•••O(4^1^) and O(5)–H•••N(1^1^), separations are 2.759, 2.792, and 2.876 Å, respectively. This confirms that the water molecules play an important role in the stabilization of the polymer.[169]

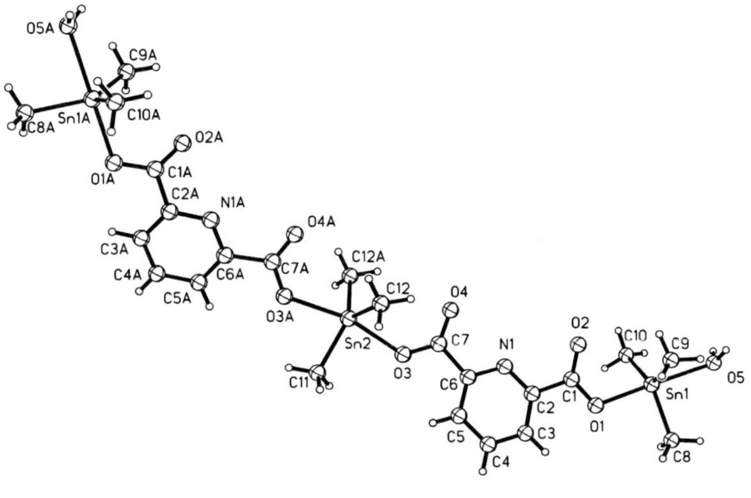

Figure 38. A diagram of the [Me$_3$Sn(dipic)]$^-$ anion. (Reproduced by permission from reference 169).

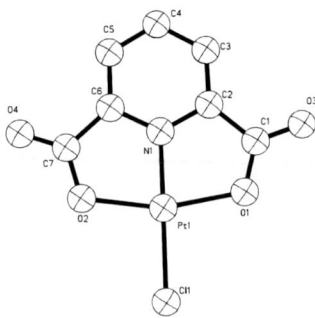

Figure 39. A diagram of the [Pt(dipic)Cl]⁻ anion. (Reproduced by permission from reference 26).

[Pt(dipic)Cl]⁻ was synthesized from dipicolinate salts and [PtCl$_4$][2-.26] Since the K$^+$ and [Bu$^n_4$N]$^+$ salts have identical spectral properties, the structure of the yellow form was characterized with the latter salt. Coexisting in the crystal of the yellow [Bu$^n_4$N][Pt(dipic)Cl].0.5H$_2$O (see Figure 39) are three slightly different types of the complex anion. The exceptionally short Pt-N distance of 1.88-1.91 Å indicated partial multiplicity.[26] The M-N distances in the dipic chelates with first-row transition metals, whose atoms are much smaller than the Pt atom, span the range of 1.88-2.17 Å.[26] The O-Pt-N "bite" angles of 80.7-81.9° indicate strain in the chelate complex. The difference of ca. 0.10 Å between the two C-O distances in the same carboxylate group indicates considerable localization of the π electrons in the exocyclic group.[26]

There was a study where two compounds were reported to be coordination polymers,[170] but in one there are some Pb(II) cations bridged by carboxylate oxygen atoms which are only 4.355(4) Å apart, so they were regarded as lying in pairs, justifying an initial description in terms of a dimeric stoichiometric unit, [Pb$_2$(dipic)$_2$(H$_2$dipic)$_2$(OH$_2$)$_6$] (Figure 40).[170]

The reaction of (NH$_4$)$_2$Ce(NO$_3$)$_6$ and CaCl$_2$ with dipicolinic acid resulted in formation of [Ca(H$_2$dipic)(OH$_2$)$_3$][Ce(dipic)$_3$].5H$_2$O.[171] Three tridentate dipic^{2-} ligands coordinate to Ce^{4+} in a slightly distorted tricapped trigonal prismatic mode (Figure 41). A tridentate dipicH$_2$ and three water molecules are bound to Ca^{2+}.[171] The protonated ligand is bound through the two carbonyl oxygen atoms and the ring nitrogen atom. Two of the dipic^{2-} ligands on Ce^{4+} act as a bridge with two different Ca^{2+} ions, thus completing a distorted square anti-prismatic type of eight-coordination around each Ca^{2+}

ion). The resulting structure is an infinite linear chain made up of alternating Ce and Ca polyhedra). The Ce-Ca distance alternates between 6.409(2) and 6.831(2) Å along the chain. One of the lattice water molecules forms a strong H-bond (HO14•••OW4 = 1.47(9) Å, OW4•••HO14•••O14 = 164(7)°) with a COOH group of dipicH$_2$. While there is no previous report of a dipic complex of Ce(IV), Na$_3$[Ce(III)(dipic)$_3$]·15H$_2$O was found[30] to be isomorphous with the Nd complex,[30] which has tricapped trigonal prismatic [Nd(dipic)$_3$]$^{3-}$ ions linked to form a three-dimensional network involving Na$^+$, carboxylate groups, and lattice water. Ca^{2+} forms a dinuclear complex [Ca(dipic)(OH$_2$)$_3$]$_2$,[33] in which Ca^{2+} is seven coordinate. A triiron linear chain is observed[23] in Fe$_3^{II}$ [(dipic)$_2$(Hdipic)$_2$(OH$_2$)$_4$], in which a central [Fe(OH)$_4$]$^{2+}$ unit is linked to two [Fe(dipicH)(dipic)]$^-$ units via carboxylate bridges.[171]

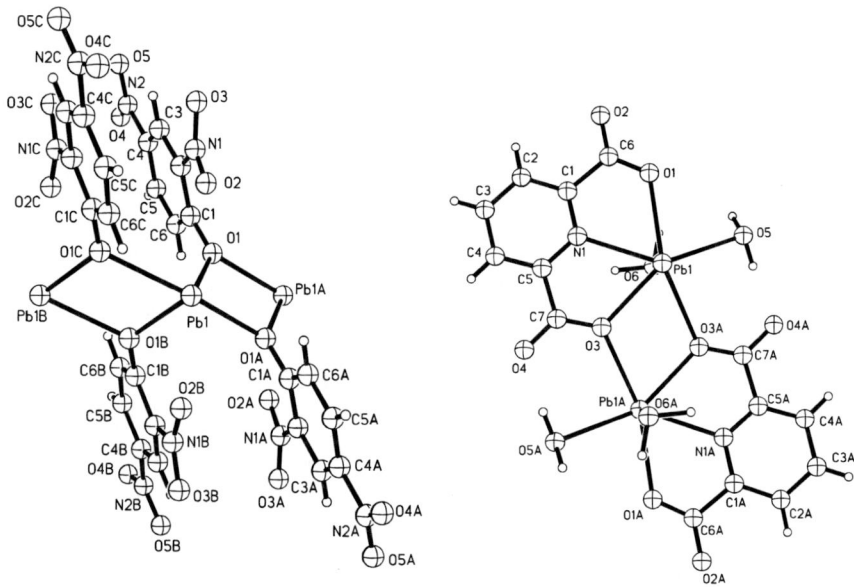

Figure 40. Diagrams of the simplest monomer and "dimer" present in [Pb$_2$(dipic)$_2$ (H$_2$dipic)$_2$(OH$_2$)$_6$]. (Reproduced by permission from reference 170).

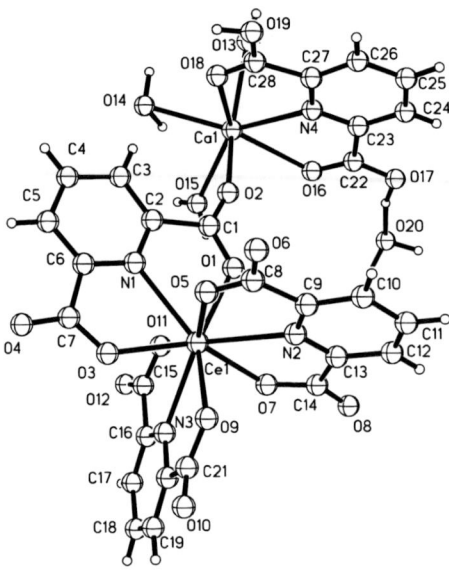

Figure 41. A diagram of [Ca(H$_2$dipic)(OH$_2$)$_3$][Ce(dipic)$_3$].5H$_2$O. (Reproduced by permission from reference 171).

The crystal structure of the newly reported [Ce(dipic)$_2$(H$_2$O)$_3$].4H$_2$O reveals an unusual hydrogen-bonded water octamer.[172] The centrosymmetric octamer is built by bridging two water molecules to the chair form of a water hexamer. The structure, predicted to be unstable relative to other octameric structures, is stabilized by hydrogen bonding with the carboxylate groups lining the cavity in the host crystal.[172] The Ce(IV) is at the center of a distorted tricapped trigonal prismatic coordination polyhedron made up of three water molecules and two dipic^{2-} ions coordinating in the tridentate chelating mode (Figure 42). The metal ligand bond distances (Å) are in the range Ce-O$_{carboxylate}$, 2.2273(1)-2.3692(1); Ce-O$_{water}$, 2.3790(2)-2.4728(2); Ce-N, 2.5171(2)-2.5218(2). The four lattice water molecules are assembled into a centrosymmetric octamer (Figure 2). The octamer has at its core a hexamer in the chair conformation with O•••O distances in the range 2.718(2)-2.788(2) Å. Two additional water molecules are attached at two diagonally opposite ends of the chair at a distance of 2.667(2) Å.[172]

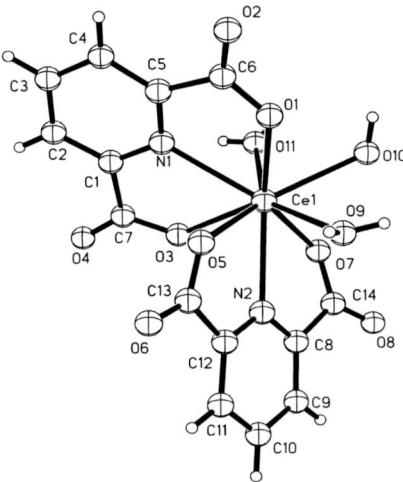

Figure 42. A diagram of [Ce(dipic)$_2$(H$_2$O)$_3$].4H$_2$O. (Reproduced by permission from reference 172).

X-ray crystallographic studies were carried out on [Gua]$_3$[Ce(dipic)$_3$].3H$_2$O and Na$_3$[Ce(dipic)$_3$].14H$_2$O.[173] The use of guanidinium counter-cation considerably reduces the amount of water molecules per elementary unit cell from 14 in the case of sodium to three.[173] At the molecular scale, no significant differences were observed in the crystal structure of the [Ce(dipic)$_3$]$^{3-}$ moieties: each isomer (Λ or Δ) presents the classical features of the tris-dipicolinate lanthanate family with small deviation from the D3 symmetry and similar average Ce–O (2.519 and 2.515 Å) and Ce–N (2.624 and 2.625 Å) distances for the Gua$^+$ and Na$^+$ complexes, respectively.[173] On the other hand, the crystal packing of both structures is very different as shown in Figure 43. In the case of the sodium derivative, the [Ce(dipic)$_3$]$^{3-}$ moieties are stacked in column, while the sodium cations are bridged by water molecules and carboxylate fragments in 1-D polymeric chains, the crystal cohesion being ensured by water molecule in a network of hydrogen bonds. On the contrary, in the case of the guanidinium compound, the crystal packing consists in alternated sheets composed by [Ce(dipic)$_3$]$^{3-}$ anions and guanidinium cations. The crystal cohesion is ensured by the guanidinum cations interconnecting two successive anionic sheets by hydrogen bonds with all the oxygen atoms of the dipicolinate fragments.[173]

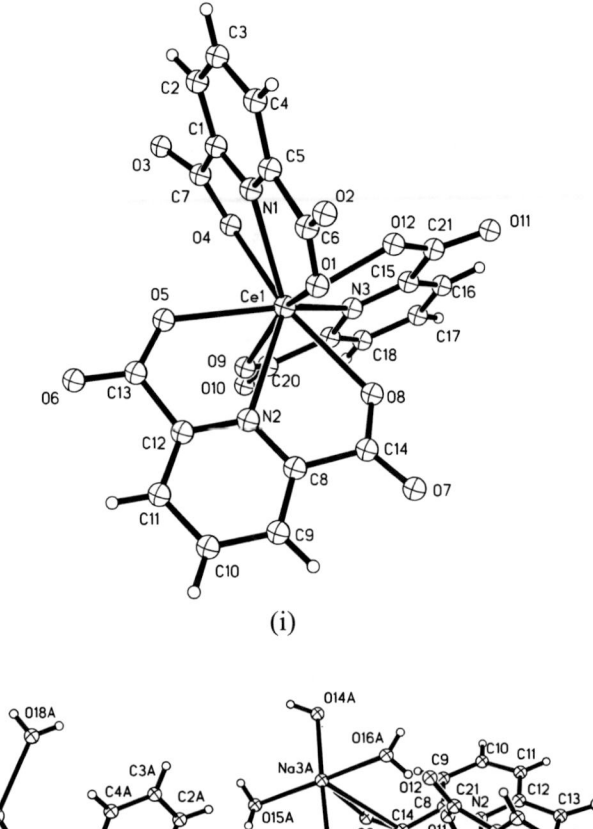

(i)

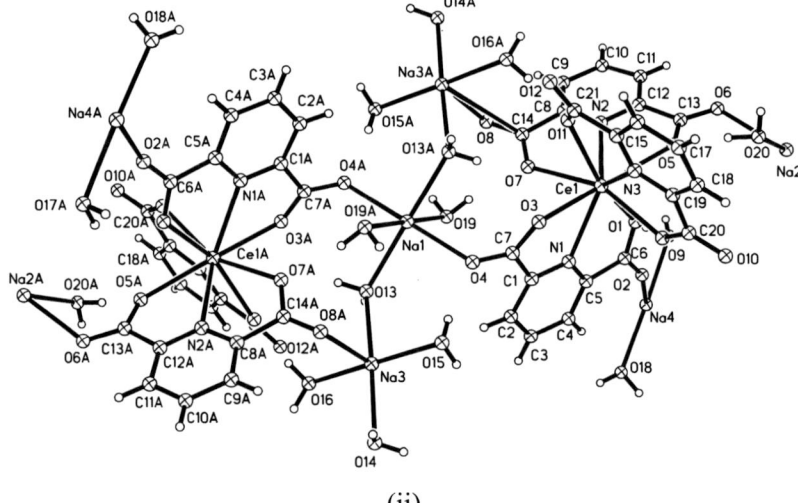

(ii)

Figure 43. Diagrams of the [Ce(dipic)$_3$]$^{3-}$ anion (i) and Na$_3$[Ce(dipic)$_3$].14H$_2$O (ii). (Reproduced by permission from reference 173).

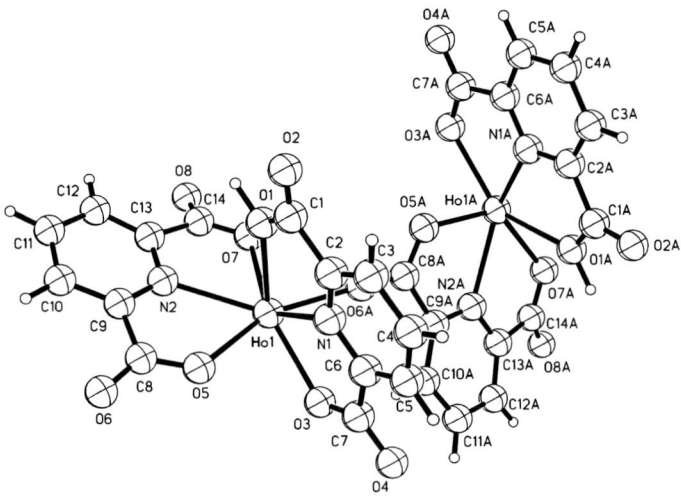

Figure 44. A diagram of [Ho(Hdipic)(dipic)]. (Reproduced by permission from reference 174).

The crystal structure of [Ho(Hdipic)(dipic)] was reported.[174] Dipic1 and dipic2 act as tridentate ligands towards the Ho^{3+} cation, with a pyridine nitrogen atom and two carboxylic oxygen atoms, (N(1), O(1), O(3)) and (N(2), O(5), O(7)), respectively (Figure 44). Moreover, dipic2 acts as a bis-monodentate ligand with two carboxylic oxygen atoms O(6) and O(8) (towards Ho^V and Ho^{IV}, respectively). Only the oxygen atoms O(2) and O(4) of dipic1 are not coordinated. One dipic anion is dianionic (dipic2), the other one (dipic1) must be monoprotonated to maintain electroneutrality. The proton was unambiguously located near O(1), from a difference Fourier synthesis. It is rather uncommon that the proton is located near an oxygen atom coordinated to the metal cation.[174]

Cs^+ and $[Co(sar)]^{3+}$ (where sar = 3,6,10,13,19-hexaazabicyclo [6.6.6] icosane] salts of the $[Eu(dipic)_3]^{3-}$ anion were chosen as materials suitable for initial structural studies because they were readily crystallized, contained an anion known usually to show strong visible luminescence.[175] The cesium and europium coordination environments are shown in Figure 45. The two cations can be considered to be bridged by the dipicolinate carboxylate groups. Two types of cesium ion, bridged by both carboxylate and water oxygen atoms, are found as part of a chain extending throughout the crystal. The chain can actually be considered as pairs of (bridged) eleven-coordinate cesium ions linked through bridging coordination to single eight coordinate cesium ions.[175] Six water molecules per formula unit of $Cs_3[Eu(dipic)_3] \cdot 9H_2O$ are involved in coordination to Cs^+, though all water molecules are involved in hydrogen bonding either to other water molecules or to the oxygen atoms of dipicolinate ligands. Although the space group is chiral and the crystal therefore optically active, the representation of the configuration of the complex anion is shown as Δ in Figure 1.[175] Metal-metal separations are of course significant with regard to electronic interactions affecting luminescence behavior; the shortest Cs•••Eu separation is 4.584 Å, while the shortest Eu•••Eu separation is 10.165 Å. Deviations of the $[Eu(dipic)_3]^{3-}$ anion from D_3 symmetry, despite the fact that the ion is situated on a twofold axis, are extremely small; the angles between the planes through the mid-point of the triangle O(11)-O(13)'-O(21)', Eu and each of the coordinated oxygens are 60.4, 59.7 and 59.9°, while the dihedral angle between planes O(11)-O(13)'-O(21)' and O(11)'-O(13)-O(21) is only 0.7°. The centroids of these planes are separated by 3.8 Å.[175]

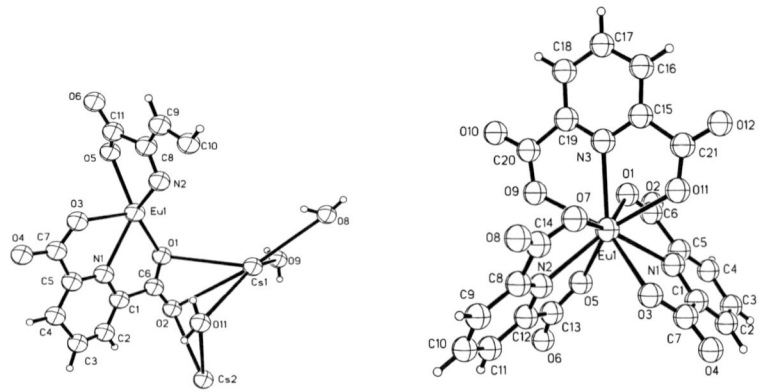

Figure 45. Diagrams of the $[Eu(dipic)_3]^{3-}$ anion. (Reproduced by permission from reference 175).

With regards to $[UO_2(Hdipic)_2] \cdot 4H_2O$ (see Figure 46), the dipicolinate units function as tridentate ligands in both cases, the uranium coordination environment being close to pentagonal bipyramidal UNO_6 in the 1:1 complex and close to hexagonal bipyramidal UN_2O_6 in the 1:2 complex.[176] The lattice of $[UO_2(Hdipic)_2] \cdot 4H_2O$ is similar to that of $[NH(C_2H_5)_3]_2[UO_2(dipic)_2] \cdot 2H_2O$.[177]

Coordination Chemistry of Dipicolinic Acid and Its Analogues 63

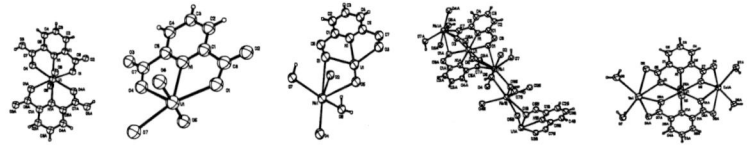

Figure 46. Diagrams of [UO$_2$(Hdipic)$_2$]·4H$_2$O. (Reproduced by permission from reference 176).

Chapter 3

CONCLUSIONS

2,6-Pyridinedicarboxylic acid (dipicolinic acid) is definitely an excellent building block in coordination and supramolecular chemistry as outlined above.

REFERENCES

[1] Froidevaux, P; Harrofield, JM; Sobolev, AN. *Inorg. Chem.*, 2000, 39, 4678-4687.
[2] Haino, T; Matsumoto, Y; Fukazawa, Y. *J. Am. Chem. Soc.*, 2005, 127, 8936-8937.
[3] Hunag, B; Prantil, MA; Gustafson, TL; Paquettte, J. R., *J. Am. Chem. Soc.*, 2003, 125, 14518-14530.
[4] Muller, G; Schmidt, B; Jiricek, J; Hopfengadner, G; Riehl, J. P; Bunzli, JCG; Piguet, C. *J. Chem. Soc., Dalton Trans.*, 2001, 2655-2662.
[5] Storm, O; Luning, U. *Eur. J. Org. Chem.*, 2003, 3109-3116.
[6] Devereux, M; McCann, M; Leon, V; McKee, V; Ball, R. J. *Polyhedron*, 2002, 21(11), 1063-1071.
[7] Kapoor, R; Kataria, A; Pathak, A; Venugopalan, P; Hundal, G; Kapoor, P. *Polyhedron*, 2005, 24 (10), 1221-1231.
[8] Kirillova, MV; Guedes da Silva, MFC; Kirillov, AM; Frausto da Silva, J.J.R; Pombeiro, A.J.L. *Inorg. Chim. Acta*, 2007, 360 (2), 506-512.
[9] Ouali, N; Bocquet, B; Rigault, S; Morgantini, P.Y; Weber, J; Piguet, C. *Inorg. Chem.*, 2002, 41 (6), 1436-1445.
[10] Renaud, F; Piguet, C; Bernardinelli, G; Bunzli, J. C. G; Hopfgartner, G. *Chem.--Eur. J*, 1997, 3 (10), 1660-1667.
[11] Renaud, F; Piguet, C; Bernardinelli, G; Bunzli, J. C. G; Hopfgartner, G. *Chem.--Eur. J*, 1997, 3(10), 1646-1659.
[12] Tse, MK; Bhor, S; Klawonn, M; Anilkumar, G; Jiao, H; Doebler, C; Spannenberg, A; Maegerlein, W; Hugl, H; Beller, M. *Chem.--Eur. J*, 2006, 12(7), 1855-1874.
[13] Jackson, A; Davis, J; Pither, RJ; Rodger, A; Hannon, M. J. *Inorg. Chem.*, 2001, 40(16), 3964-3973.

[14] Takusagawa, F; Hirotsu, K; Shimada, A. *Bull. Chem. Soc. Jpn.* 1973, 46(7), 2020-2027.
[15] Moghimi, A; Sharif, M. A; Aghabozorg, H. *Acta Crystallgr*, 2004, E60(10), o1790-o1792.
[16] Smith, G; White, JM. *Aust. J. Chem.*, 2001, 54, 97-100.
[17] Ramsay, W., *Jahresber. Fortschr. Chem.*, 1877, 437.
[18] Udo, S. *Nippon Nogei Kagaku Kaishi*, 1936, 12, 386-394.
[19] Laine, P; Gourdon, A; Launay, JP. *Inorg. Chem.*, 1995, 34(21), 5129-5137.
[20] Ventur, D; Wieghardt, K; Weiss, J. *Z. Anorg. Allg. Chem.*, 1985, 524, 40-50.
[21] Drew, MGB; Fowles, GWA; Matthews, RW; Walton, RA. *J. Am. Chem., Soc*, 1969, 91(27), 7769-7771.
[22] Drew, MGB; Matthews, RW; Walton, RA. *J. Chem. Soc. A*, 1970, (9) 1405-1410.
[23] Laine, P; Gourdon, A; Launay, JP. *Inorg. Chem.*, 1995, 34(21), 5138-5149.
[24] Laine, P; Gourdon, A; Launay, JP. *Inorg. Chem.*, 1995, 34(21), 5156-5165.
[25] Laine, P; Gourdon, A; Launay, JP; Tuchagues, JP. *Inorg. Chem.*, 1995, 34(21), 5150-5155.
[26] Zhou, XY; Kostic, NM. *Inorg. Chem.*, 1988, 27(24), 4402-4408.
[27] Chessa, G; Marangoni, G; Pitteri, B; Bertolasi, V; Gilli, G; Ferretti, V. *Inorg. Chim. Acta*, 1991, 185(2), 201-210.
[28] Espinet, P; Miguel, JA; Garcia-Granda, S; Miguel, D. *Inorg. Chem.*, 1996, 35(8), 2287-2291.
[29] Albertsson, J. *Acta Chem. Scand*, 1970, 24(4), 1213-1229.
[30] Albertsson, J. *Acta Chem. Scand*, 1972, 26(3), 1023-1044.
[31] Albertsson, J. *Acta Chem. Scand*, 1972, 26(3), 985-1004.
[32] Palmer, KJ; Wong, RY; Lewis, JC. *Acta Crystallogr., Sect. B*, 1972, 28, (Pt. 1), 223-228.
[33] Strahs, G; Dickerson, RE. *Acta Crystallogr., Sect. B*, 1968, 24(4), 571-578.
[34] Groves, JT; Kady, I. O. *Inorg. Chem.*, 1993, 32(18), 3868-3872.
[35] Harrington, PC; Wilkins, RG. *J. Inorg. Biochem*, 1980, 12(2), 107-118.
[36] Holder, AA; Brown, RFG; Marshall, SC; Payne, VCR; Cozier, MD; Alleyne, WA., Jr; Bovell, CO. *Transition Met. Chem.*, 2000, 25(5), 605-611.

[37] Mauk, AG; Bordignon, E; Gray, HB. *J. Am. Chem. Soc.*, 1982, 104(26), 7654-7657.
[38] Mauk, AG; Coyle, CL; Bordignon, E; Gray, HB. *J. Am. Chem. Soc.*, 1979, 101(17), 5054-5056.
[39] Varey, JE; Lamprecht, GJ; Fedin, VP; Holder, A; Clegg, W; Elsegood, MRJ; Sykes, AG. *Inorg. Chem.*, 1996, 35(19), 5525-5530.
[40] Balavoine, G; Barton, DHR; Gref, A; Lellouche, I. *Tetrahedron*, 1992, 48(10), 1883-1894.
[41] Cofre, P; Richert, SA; Sobkowiak, A; Sawyer, DT. *Inorg. Chem.*, 1990, 29(14), 2645-2651.
[42] Dalton, H. *Adv. Applied Microbiol*, 1980, 26, 71-87.
[43] Ericson, A; Hedman, B; Hodgson, KO; Green, J; Dalton, H; Bentsen, J. G; Beer, RH; Lippard, SJ. *J. Am. Chem. Soc.*, 1988, 110(7), 2330-2332.
[44] Fox, BG; Surerus, KK; Munck, E; Lipscomb, JD. *J. Biol. Chem.*, 1988, 263(22), 10553-10556.
[45] Sheu, C; Sawyer, DT. *J. Am. Chem. Soc.*, 1990, 112(22), 8212-8214.
[46] Sheu, C; Sobkowiak, A; Jeon, S; Sawyer, DT. *J. Am. Chem. Soc.*, 1990, 112(2), 879-881.
[47] Sugimoto, H; Tung, HC; Sawyer, DT. *J. Am. Chem. Soc*, 1988, 110(8), 2465-2470.
[48] Tung, HC; Kang, C; Sawyer, DT. *J. Am. Chem. Soc.*, 1992, 114(9), 3445-3455.
[49] Chiesi Villa, A; Guastini, C; Musatti, A; Nardelli, M. *Gazz. Chim. Ital.*, 1972, 102(3), 226-233.
[50] Gaw, H; Robinson, WR; Walton, RA. *Inorg. Nucl. Chem. Lett*, 1971, 7(8), 695-699.
[51] Hakansson, K; Lindahl, M; Svensson, G; Albertsson, J. *Acta Chem. Scand*, 1993, 47, 449-455.
[52] Okabe, N; Oya, N. *Acta Crystallogr., Sect. C: Cryst. Struct. Commun*, 2000, C56(3), 305-307.
[53] Quaglieri, P; Loiseleur, H; Thomas, G. *Acta Crystallogr., Sect. B*, 1972, 28(Pt. 8), 2583-90.
[54] Biagini Cingi, M; Chiesi Villa, A; Guastini, C; Nardelli, M. *Gazz. Chim. Ital*, 1971, 101(11), 825-832.
[55] Biagini Cingi, M; Chiesi Villa, A; Guastini, C; Nardelli, M. *Gazz. Chim. Ital*, 1972, 102(11), 1026-1033.
[56] Yang, L; Crans, DC; Miller, SM; la Cour, A; Anderson, OP; Kaszynski, PM; Godzala, ME; Austin, LD; Willsky, GR. *Inorg. Chem.*, 2002, 41(19), 4859-4871.

[57] MacDonald, JC; Dorrestein, PC; Pilley, MM; Foote, MM; Lundburg, JL; Henning, RW; Schultz, AJ; Manson, JL. *J. Am. Chem. Soc.*, 2000, 122(47), 11692-11702.
[58] Browning, K; Abboud, KA; Palenik, GJ. *J. Chem. Crystallogr*, 1995, 25(12), 851-855.
[59] Carmona, P. *Spectrochim. Acta Part A: Molecular Spectroscopy*, 1980, 36(7), 705-712.
[60] Starosta, W; Ptasiewicz-Bak, H; Leciejewicz, J. *J. Coord. Chem.*, 2002, 55(8), 873-881.
[61] Soleimannejad, J; Aghabozorg, H; Hooshmand, S; Adams, H. *Acta Crystallogr., Section E*, 2007, E63, (12), m3089-m3090, m3089/1-m3089/14.
[62] Gaetjens, J; Meier, B; Adachi, Y; Sakurai, H; Rehder, D. *Eur. J. Inorg. Chem.*, 2006, (18), 3575-3585.
[63] Manohar, H; Schwarzenbach, D. *Helv. Chim. Acta*, 1974, 57(4), 1086-1095.
[64] Leik, R; Zsolnai, L; Huttner, G; Neuse, EW; Brintzinger, HH. *J. Organomet. Chem.*, 1986, 312(2), 177-182.
[65] Anderson, SJ; Brown, DS; Norbury, AH., *J. Chem. Soc., Chem. Commun.*, 1974, 996.
[66] Sanner, RD; Dugga, DM; Mckenzie, TC; Marsh, RE; Bercaw, JE. *J. Am. Chem. Soc.*, 1976, 98, 8358-8365.
[67] Zeinstra, JD; Teuben, JH; Jellinek, F. *J. Organomet. Chem.* 1979, 170, 39-50.
[68] Besancon, J; Top, S; Tirouflet, J; Dusausoy, Y; Lecomte, C; Protas, J. *J. Organomet. Chem.*, 1977, 127, 153-168.
[69] Curtis, MD; Thanedar, S; Butler, WM. *Organometallics*, 1984, 3, 1855-1859.
[70] Huffman, JC; Moloy, kG; Marsella, JA; Caulton, KG. *J. Am. Chem. Soc*, 1980, 102, 3009-3014.
[71] Silver, ME; Eisenstein, O; Fay, RC. *Inorg. Chem.*, 1983, 22, 759-770.
[72] Thewalt, U; Lasser, W. *J. Organomet. Chem.*, 1984, 276, 341-347.
[73] Schwarzenbach, D. *Helv. Chim. Acta*, 1972, 55(8), 2990-3004.
[74] Smee, JJ; Epps, JA; Teissedre, G; Maes, M; Harding, N; Yang, L; Baruah, B; Miller, SM; Anderson, OP; Willsky, GR; Crans, DC. *Inorg. Chem.*, 2007, 46(23), 9827-9840.
[75] Smee, JJ; Epps, JA; Ooms, K; Bolte, SE; Polenova, T; Baruah, B; Yang, L; Ding, W; Li, M; Willsky, GR; Cour, Al; Anderson, OP; Crans, DC. *J. Inorg. Biochem*, 2009, 103(4), 575-584.

[76] Cocco, MT; Onnis, V; Ponticelli, G; Meier, B; Rehder, D; Garribba, E; Micera, G. *J. Inorg. Biochem*, 2007, 101(1), 19-29.
[77] Parajón-Costa, BS; Piro, OE; Pis-Diez, R; Castellano, EE; González-Baró, AC. *Polyhedron*, 2006, 25(15), 2920-2928.
[78] Kavitha, SJ; Panchanatheswaran, K; Elsegood, MRJ; Dale, SH. *Inorg. Chim. Acta*, 2006, 359(4), 1314-1320.
[79] Yin-Zhuang, Z; Jinli, L. *Inorg. Chem. Commun*, 2009, 12(3), 243-245.
[80] Gonzalez-Baró, AC; Castellano, EE; Piro, OE; Parajón-Costa, BS. *Polyhedron*, 2005, 24(1), 49-55.
[81] Crans, DC; Mahroof-Tahir, M; Johnson, MD; Wilkins, PC; Yang, L; Robbins, K; Johnson, A; Alfano, JA; Godzala, ME; Austin, LT; Willsky, GR. *Inorg. Chim. Acta*, 2003, 356, 365-378.
[82] Bersted, BH; Belford, RL; Paul, IC. *Inorg. Chem.*, 1968, 7(8), 1557-1562.
[83] Chatterjee, M; Ghosh, S; Wu, BM; Mak, TCW. *Polyhedron*, 1998, 17(8), 1369-1374.
[84] Crans, DC; Yang, L; Jakusch, T; Kiss, T. *Inorg. Chem.*, 2000, 39(20), 4409-4416.
[85] Xing, YH; Aoki, K; Bai, FY. *J. Coord. Chem.*, 2004, 57(2), 157-165.
[86] Bersted, BH. Crystal, molecular, and electronic structure of vanadyl(IV) 2,6-lutidinate tetrahydrate. 1969, 94 CAN 74:104184 AN 1971:104184.
[87] Hartkamp, H. *Angew. Chem.*, 1959, 71, 553.
[88] Thompson, KH; Orvig, C. *Coord. Chem. Rev.*, 2001, 219-221, 1033-1053.
[89] Nuber, B; Weiss, J. *Acta Crystallogr., Sect. B*, 1981, B37(4), 947-948.
[90] Holder, AA; VanDerveer, D. *Acta Crystallogr., Sect. E: Struct. Rep. Online*, 2007, E63(8), m2051-m2052.
[91] Casny, M; Rehder, D. *Chem. Commun.*, 2001, (10), 921-922.
[92] Shaver, A; Ng, JB; Hall, DA; Lum, BS; Posner, BI. *Inorg. Chem.*, 1993, 32(14), 3109-3113.
[93] Shaver, A; Hall, DA; Ng, JB; Lebuis, AM; Hynes, RC; Posner, BI. *Inorg. Chim. Acta*, 1995, 229(1-2), 253-260.
[94] Kiss, E; Benyei, A; Kiss, T. *Polyhedron*, 2003, 22(1), 27-33.
[95] Chatterjee, M; Ghosh, S; Nandi, AK. *Polyhedron*, 1997, 16(17), 2917-2923.
[96] Farrugia, LJ. *J. Appl. Crystallogr*, 1997, 30(5, Pt. 1), 565.
[97] Cornman, CR; Kampf, J; Lah, MS; Pecoraro, VL. *Inorg. Chem.*, 1992, 31(11), 2035-2043.

[98] Cotton, FA; Czuchajowska, J; Feng, X. *Inorg. Chem.*, 1991, 30(2), 349-353.
[99] Yang, L; la Cour, A; Anderson, OP; Crans, DC. *Inorg. Chem.*, 2002, 41(24), 6322-6331.
[100] Nuber, B; Weiss, J; Wieghardt, K. *Z. Naturforsch., B: Anorg. Chem., Org. Chem*, 1978, 33B(3), 265-267.
[101] Li, X; Lah, MS; Pecoraro, VL. *Inorg. Chem.*, 1988, 27(25), 4657-4664.
[102] Cotton, FA; Day, VW; Hazen, EE., Jr; Larsen, S. *J. Am. Chem. Soc.*, 1973, 95(15), 4834-4840.
[103] Katrusiak, A; Szafranski, M. *J. Mol. Struct*, 1996, 378(3), 205-223.
[104] Russell, VA; Etter, MC; Ward, MD. *J. Am. Chem. Soc.*, 1994, 116(5), 1941-1952.
[105] Smee Jason, J; Epps Jason, A; Teissedre, G; Maes, M; Harding, N; Yang, L; Baruah, B; Miller Susie, M; Anderson Oren, P; Willsky Gail, R; Crans Debbie, C. *Inorg. Chem.*, 2007, 46(23), 9827-9840.
[106] Addison, AW; Rao, TN; Reedijk, J; Van Rijn, J; Verschoor, GC. *J. Chem. Soc., Dalton Trans.*, 1984, (7), 1349-1356.
[107] Drew, RE; Einstein, FWB. *Inorg. Chem.*, 1973, 12(4), 829-385.
[108] Mimoun, H; Chaumette, P; Mignard, M; Saussine, L; Fischer, J; Weiss, R. *Nouveau Journal de Chimie*, 1983, 7(8-9), 467-475.
[109] Tinant, B; Bayot, D; Devillers, M. *Z. Kristallogr.- New Cryst. Struct*, 2003, 218(4), 477-478.
[110] Hon, PK; Belford, RL; Pfluger, CE. *J. Chem. Phys.*, 1965, 43(4), 1323-1333.
[111] Amin, SS; Cryer, K; Zhang, B; Dutta, SK; Eaton, SS; Anderson, OP; Miller, SM; Reul, BA; Brichard, SM; Crans, DC. *Inorg. Chem.*, 2000, 39(3), 406-416.
[112] Crans, DC; Keramidas, AD; Mahroof-Tahir, M; Anderson, OP; Miller, MM. *Inorg. Chem.*, 1996, 35(12), 3599-3606.
[113] Scheidt, WR; Countryman, R; Hoard, JL., *J. Am. Chem. Soc.*, 1971, 93(16), 3878-3882.
[114] Melchior, M; Thompson, KH; Jong, JM; Rettig, SJ; Shuter, E; Yuen, VG; Zhou, Y; McNeill, JH; Orvig, C. *Inorg. Chem.*, 1999, 38(10), 2288-2293.
[115] Renolds, JG; Sendlinger, SC; Murray, AM; Hoffman, JC; Christou, G., *Inorg. Chem.*, 1995, 34, 5745-5752.
[116] Dewey, TM; Du Bois, J; Raymond, KN. *Inorg. Chem.*, 1993, 32(9), 1729-1738.
[117] Taylor, R; Kennard, O; Versichel, W., *Acta Crystallogr*, 1984, B40, 280.

[118] Kojima, A; Okazaki, K; Ooi, S; Saito, K. *Inorg. Chem.*, 1983, 22(8), 1168-1174.
[119] Payne, VCR; Headley, OSC; Stibrany, RT; Maragh, PT; Dasgupta, TP; Newton, AM; Holder, AA. *J. Chem. Crystallogr*, 2007, 37(4), 309-314.
[120] Fuerst, W; Gouzerh, P; Jeannin, Y. *J. Coord. Chem.*, 1979, 8(4), 237-243.
[121] Johnson, JE; Jacobson, RA. *Acta Crystallogr., Sect. B*, 1973, 29(8), 1669-1674.
[122] Janiak, C. *Dalton Trans.*, 2000, (21), 3885-3896.
[123] González-Baró, AC; Pis-Diez, R; Piro, OE; Parajón-Costa, BS. *Polyhedron*, 2008, 27(2), 502-512.
[124] Devereux, M; McCann, M; Leon, V; McKee, V; Ball, RJ., *Polyhedron*, 2002, 21(11), 1063-1071.
[125] Park, H; Lough, AJ; Kim, JC; Jeong, MH; Kang, YS. *Inorg. Chim. Acta*, 2007, 360(8), 2819-2823.
[126] Aghabozorg, H; Mohamad Panah, F; Sadr-Khanlou, E. *Analyt. Sciences: X-Ray Structure Analysis Online*, 2007, 23(7), x139-x140.
[127] Aghabozorg, H; Nemati, A; Derikvand, Z; Ghadermazi, M. *Acta Crystallogr., Section E: Structure Reports Online*, 2007, E63(12), m2921, m2921/1-m2921/15.
[128] Aghabozorg, H; Sadrkhanlou, E; Soleimannejad, J; Adams, H. *Acta Crystallogr., Section E: Structure Reports Online*, 2007, E63(6), m1760.
[129] Cousson, A; Nectoux, F; Robert, F; Rizkalla, EN. *Acta Crystallogr., Section C: Crystal Structure Communications*, 1995, C51(5), 838-840.
[130] Rafizadeh, M; Mehrabi, B; Amani, V. *Acta Crystallogr., Section E: Structure Reports Online*, 2006, E62(6), m1332-m1334.
[131] Sanyal, GS; Ganguly, R; Nath, PK; Butcher, RJ. *J. Indian Chem. Soc.*, 2002, 79(6), 489-491.
[132] Zhao, QH; Zhang, MS; Fang, RB. *Acta Crystallogr., Section E: Structure Reports Online*, 2005, E61(12), m2575-m2577.
[133] Uçar, I; Bulut, A; Karadag, A; Kazak, C. *J. Mol. Struct*, 2007, 837(1-3), 38-42.
[134] Braga, D; Bazzi, C; Maini, L; Grepioni, F. *CrystEngComm*, 1999, No Given, Article 5.
[135] Du, M; Cai, H; Zhao, XJ. *Inorg. Chim. Acta*, 2006, 359(2), 673-679.
[136] Liu, FC; Ouyang, J. *Acta Crystallogr., Section E: Structure Reports Online*, 2007, E63(10), m2557, Sm2557/1-Sm2557/7.
[137] Su, H; Wen, YH; Feng, Y. L. *Zeitschrift fuer Kristallographie - New Crystal Structures*, 2005, 220(4), 560-562.

[138] Sun, Q; Gao, Q; Zhang, W; Song, Y; Xu, Z; Su, B; Zhao, J. *J. Coord. Chem*, 2008, 61(5), 669-676.
[139] Wang, L; Duan, L; Wang, E; Xiao, D; Li, Y; Lan, Y; Xu, L; Hu, C. *Transition Met. Chem.*, 2004, 29(2), 212-215.
[140] Bedetti, R; Biader Ceipidor, U; Carunchio, V; Tomassetti, M. *Annali di Chimica*, 1976, 66(11-12), 741-752.
[141] Moody, L; Balof, S; Smith, S; Rambaran, VH; VanDerveer, D; Holder, A. A. *Acta Crystallogr., Sect. E: Struct. Rep. Online*, 2008, E64(1), m262-m263, m262/1-m262/14.
[142] Alcock, NW; Clarkson, GJ; Lawrance, GA; Moore, P. *Aust. J. Chem.*, 2004, 57(6), 565-570.
[143] Jiang, YM; Yin, ZJ., *Wuji Huaxue Xuebao*, 2001, 17, 589.
[144] Sileo, EE; Blesa, MA; Rigotti, G; Rivero, BE; Castellano, EE. *Polyhedron*, 1996, 15(24), 4531-4540.
[145] Brandi-Blanco, MP; Choquesillo-Lazarte, D; GarcIa-Collado, CG; González-Pérez, JM; Castiñeiras, A; Niclós-Gutiérrez, J. *Inorg. Chem. Commun.*, 2005, 8(2), 231-234.
[146] Su, CC; Chiu, SY. *Polyhedron*, 1996, 15(15), 2623-2631.
[147] Bresciani-Pahor, N; Nardin, G; Bonomo, RP; Rizzarelli, E. *J. Chem. Soc., Dalton Trans.*, 1984, (12), 2625-3260.
[148] Nardin, G; Randaccio, L; Bonomo, RP; Rizzarelli, E. *J. Chem. Soc., Dalton Trans.*, 1980, (3), 369-375.
[149] Gao, HL; Yi, L; Zhao, B; Zhao, XQ; Cheng, P; Liao, DZ; Yan, SP. *Inorg. Chem.*, 2006, 45(15), 5980-5988.
[150] Prasad, TK; Rajasekharan, MV. *Polyhedron*, 2007, 26(7), 1364-1372.
[151] Prasad, TK; Rajasekharan, MV. *Inorg. Chem. Commun*, 2005, 8(12), 1116-1119.
[152] Prasad, TK; Sailaja, S; Rajasekharan, MV. *Polyhedron*, 2005, 24(12), 1487-1496.
[153] Sailaja, S; Rajasekharan, MV. *Acta Crystallogr., Sect. E: Struct. Rep. Online*, 2001, E57, (8), m341-m343.
[154] Swarnabala, G; Rajasekharan, MV. *Inorg. Chem.*, 1998, 37(7), 1483-1485.
[155] Shen, E; Lue, J; Li, Y; Wang, E; Hu, C; Xu, L. *J. Solid State Chem.*, 2004, 177(11), 4372-4376.
[156] Arnáiz, FJ; Aguado, R; Pedrosa, MR; De Cian, A; Fischer, J. *Polyhedron*, 2000, 19(20-21), 2141-2147.
[157] Sukanya, D; Prabhakaran, R; Natarajan, K. *Polyhedron*, 2006, 25(11), 2223-2228.

[158] Annibale, G; Cattalini, L; Canovese, L; Pitteri, B; Tiripicchio, A; Tiripicchio, CM; Tobe, ML. *J. Chem. Soc., Dalton Trans.*, 1986, 1101-1105.
[159] Espinet, P; Lorenzo, C; Miguel, JA; Bois, C; Jeannin, Y. *Inorg. Chem.*, 1994, 33, 2052-2055.
[160] Yamamoto, JH; Yoshida, W; Jensen, CM. *Inorg. Chem.*, 1991, 30, 1353-1357.
[161] Xue, L; Che, YX; Zheng, JM. *J. Coord. Chem.*, 2007, 60(13), 1381-1386.
[162] Allen, FH; Davies, JE; Galloy, JJ; Johnson, O; Kennard, O; Macrae, C. F; Mitchell, EM; Mitchell, GF; Smith, JM; Watson, DG. *J. Chem. Inf. Comput. Sci.*, 1991, 31(2), 187-204.
[163] MacDonald, JC; Luo, TJM; Palmore, GTR. *Cryst. Growth Des.*, 2004, 4(6), 1203-1209.
[164] Sileo, EE; Blesa, MA; Rigotti, G; Rivero, BE; Castellano, EE. *Polyhedron*, 1996, 15(24), 4531-4540.
[165] Moghimi, A; Sheshmani, S; Shokrollahi, A; Aghabozorg, H; Shamsipur, M; Kickelbick, G; Aragoni, M. C; Lippolis, V. *Zeit. fuer Anorg. und Allgem. Chemie*, 2004, 630(4), 617-624.
[166] Moghimi, A; Ranjbar, M; Aghabozorg, H; Jalali, F; Shamsipur, M; Chadah, RK. *J. Chem. Res., Synop.*, 2002, (10), 477-479, 1047-1065.
[167] Moghimi, A; Ranjbar, M; Aghabozorg, H; Jalali, F; Shamsipur, M; Chadha, KK. *Can. J. Chem.*, 2002, 80(12), 1687-1696.
[168] Ranjbar, M; Aghabozorg, H; Moghimi, A., *Acta Crystallogr*, 2002, E58, m304.
[169] Ma, C; Li, J; Zhang, R; Wang, D. *J. Organomet. Chem.*, 2006, 691(8), 1713-1721.
[170] Harrowfield, JM; Shahverdizadeh, GH; Soudi, AA. *Supramolecular Chemistry*, 2003, 15(5), 367-373.
[171] Swarnabala, G; Rajasekharan, MV. *Inorg. Chem.*, 1998, 37(7), 1483-1485.
[172] Prasad, TK; Rajasekharan, MV. *Cryst. Growth Des.*, 2006, 6(2), 488-491.
[173] D'Aléo, A; Toupet, L; Rigaut, S; Andraud, C; Maury, O. *Optical Mat.* 2008, 30(11), 1682-1688.
[174] Fernandes, A; Jaud, J; Dexpert-Ghys, J; Brouca-Cabarrecq, C. *Polyhedron*, 2001, 20(18), 2385-2391.
[175] Brayshaw, PA; Buenzli, JCG; Froidevaux, P; Harrowfield, JM; Kim, Y; Sobolev, AN. *Inorg. Chem.*, 1995, 34(8), 2068-2076.

[176] Harrowfield, JM; Lugan, N; Shahverdizadeh, GH; Soudi, AA; Thuery, P. *Eur. J. Inorg. Chem.*, 2006, (2), 389-396.
[177] Masci, B; Thuery, P. *Polyhedron*, 2005, 24(2), 229-237.

INDEX

A

accounting, 22
acetonitrile, 50
acid, vii, 1, 3, 4, 5, 6, 8, 10, 13, 26, 31, 33, 35, 40, 44, 49, 50, 53, 55, 56, 65
acidic, 34
alkaline earth metals, 5
amine, 27, 28, 29, 40, 43
amino, 18, 19
ammonium, 8
ammonium salts, 8
aromatic rings, 42, 51
asymmetry, 9, 15, 17, 20
atoms, vii, 1, 6, 7, 8, 9, 10, 12, 13, 14, 15, 16, 18, 20, 22, 25, 26, 27, 30, 34, 35, 38, 40, 42, 43, 44, 45, 49, 51, 53, 54, 55, 56, 59, 61

B

base, 53
basicity, 46
bonding, 1, 14, 15, 17, 20, 22, 25, 26, 28, 32, 34, 36, 38, 39, 40, 52, 53, 54, 55, 58, 62
bonds, 4, 6, 9, 12, 13, 14, 18, 19, 20, 22, 25, 34, 37, 42, 43, 44, 45, 49, 54, 59
Butcher, 73

C

Ca^{2+}, 56
cadmium, 53
carboxyl, 10, 12, 55
carboxylic acid, 2
carboxylic groups, 2, 17, 45
cation, 9, 15, 16, 18, 27, 29, 40, 41, 45, 54, 59, 61
C-C, 22
cesium, 61
chelates, 10, 56
chlorine, 30
chromium, 27
cleavage, 3
clusters, 46
C-N, 54
CO2, 7
cobalt, 35, 38
compensation, 41
composition, vii, 1
compounds, 45, 56
compression, 34
configuration, 25, 62
construction, 53
COOH, 57
coordination, vii, 1, 3, 4, 5, 6, 8, 9, 14, 15, 18, 19, 22, 25, 27, 30, 34, 35, 37, 38, 39,

40, 42, 43, 44, 45, 46, 50, 53, 54, 55, 56, 58, 61, 62, 65
copper, 36, 39, 40, 42, 43, 44, 45
covalent bond, 46
covalent bonding, 46
creatinine, 1, 16
crystal structure, vii, 1, 5, 27, 30, 31, 32, 35, 36, 38, 41, 46, 53, 58, 59, 61
crystalline, 4
crystallization, 9, 12, 33, 35
crystals, 9, 34, 45
cycles, 51

D

derivatives, vii, 1, 5, 6, 18, 51
deviation, 7, 8, 13, 16, 30, 45, 50, 53, 59
dimerization, 51
Dipicolinic acid, vii, 3
disorder, 14, 22
distortions, 40
DMF, 46, 47, 48
DNA, 3
donors, 20, 39, 42
double bonds, 8
drawing, 16

E

electron, 3, 11, 19, 20, 27, 37, 54
electronic structure, 71
electrons, 56
elongation, 34, 40, 45
environment, 8, 13, 16, 25, 30, 51, 62
ester, vii, 1
ethanol, 42
europium, 61

F

families, 51
flexibility, 43
force, 41
formation, 15, 16, 39, 41, 53, 56

formula, vii, 1, 14, 15, 36, 62
fragments, 1, 54, 59

G

geometrical parameters, 17
geometry, 7, 8, 10, 18, 19, 22, 25, 27, 32, 34, 38, 39, 40, 46, 55

H

H-bonding, 17, 42
host, 58
hydrogen, 1, 4, 9, 12, 13, 14, 15, 18, 20, 22, 26, 27, 32, 34, 35, 36, 38, 39, 40, 52, 53, 54, 55, 58, 59, 62
hydrogen atoms, 9, 37
hydrogen bonds, 4, 12, 13, 14, 15, 18, 22, 35, 36, 38, 39, 55, 59
hydroxyl, 14, 15, 19

I

identification, 36
in transition, 3
intermolecular interactions, 1
inversion, 17, 28, 33, 35
ions, vii, 1, 9, 14, 16, 18, 20, 32, 35, 38, 40, 45, 51, 56, 58, 61

K

K^+, 26, 56

L

labeling, 25, 35
lead, 53
ligand, 1, 3, 4, 6, 7, 8, 9, 10, 13, 14, 15, 16, 18, 19, 22, 25, 30, 32, 35, 36, 38, 40, 41, 42, 43, 45, 46, 49, 50, 51, 53, 56, 58, 61
localization, 2, 17, 27, 56

luminescence, 61
Luo, 75
lying, 1, 8, 17, 56

M

magnesium, 5
majority, 19
mass, vii, 1
materials, 4, 61
metal complexes, vii, 54
metal ion, vii, 1, 4, 8, 30, 32, 53
metal ions, vii, 1, 4
metals, 5, 35
methanol, 39, 40
methyl group, 22
methyl groups, 22
molecular structure, 35, 43, 53
molecules, 10, 12, 14, 15, 18, 22, 26, 27, 30, 32, 35, 36, 38, 39, 42, 45, 54, 55, 56, 58, 59, 62
molybdenum, 46
monomers, 51
motif, 52

N

Na^+, 15, 25, 57, 59
Nd, 57
neutral, 2, 35, 36, 41, 43
NH2, 8, 9, 16, 18, 19, 20, 21
nickel, 36, 39
nitrogen, 1, 5, 8, 9, 10, 11, 13, 14, 16, 19, 22, 25, 26, 30, 35, 45, 49, 50, 53, 54, 56, 61
NMR, 51
nonlinear optics, 3
nucleus, 43

O

OH, 14, 15, 16, 19, 20, 21, 23, 25, 33, 42, 57
overlap, 22

oxidation, 1, 8, 25
oxygen, 1, 4, 8, 9, 10, 11, 12, 13, 14, 15, 16, 18, 20, 22, 25, 26, 28, 30, 33, 34, 35, 37, 38, 39, 40, 45, 46, 50, 53, 56, 59, 61

P

pairing, 1, 53
palladium, 50, 51
parallel, 9, 12, 18, 32
permission, 2, 6, 7, 10, 11, 14, 16, 17, 19, 20, 23, 27, 28, 29, 31, 32, 33, 36, 37, 38, 39, 41, 42, 44, 45, 46, 48, 49, 50, 52, 53, 55, 56, 57, 58, 59, 60, 61, 62, 63
peroxide, 5
pH, 34
phenol, 26, 32
phosphate, 9, 18
phosphorus, 50
physical properties, 3
platinum, 51
PM, 69
polymer, 39, 41, 55
polymeric chains, 15, 59
polymers, 44, 55, 56
potassium, 10, 20, 26
probability, 28, 29, 40

R

reactants, 34
reactions, 50
relaxation, 34
requirements, 45
rings, 10, 27, 40, 52
rubidium, 27
ruthenium, 49

S

salts, 4, 25, 44, 56, 61
self-assembly, 44, 46
showing, 19, 42, 49
silver, 36

skeleton, 17
sodium, 5, 15, 18, 30, 35, 59
solid state, 3, 53
solution, 26
species, 30, 34
spectroscopy, 51
stabilization, 55
stable complexes, 1
standard deviation, 5
state, 25
states, 1, 8
stoichiometry, 9, 16
strontium, 5
structure, 1, 5, 6, 9, 14, 15, 17, 18, 19, 20, 25, 26, 27, 32, 33, 34, 35, 36, 38, 39, 40, 41, 43, 45, 46, 50, 52, 54, 56, 57, 58
substitution, 19, 54
Sun, 74
symmetry, 7, 13, 15, 18, 19, 20, 22, 40, 51, 54, 59, 62
synthesis, 4, 35, 46, 61

T

techniques, 25
tin, 54, 55
titanium, 5, 7
topology, 52
torsion, 50, 53

transformation, 13
transition metal, 3, 4, 5, 6, 56
transition metal ions, 3

U

uranium, 62

V

vanadium, 8, 10, 11, 12, 13, 14, 15, 16, 18, 19, 21, 22, 24, 25, 26

W

water, 1, 5, 9, 10, 12, 14, 15, 18, 22, 26, 27, 30, 32, 33, 34, 35, 36, 38, 39, 42, 45, 53, 54, 55, 56, 58, 59, 61
weak interaction, 17

X

X-ray crystallography, 13, 14, 18, 25, 35, 54
X-ray diffraction, 46, 54